OORT AND THE UNIVERSE

JAN HENDRIK OORT

OORT AND THE UNIVERSE

A Sketch of Oort's Research and Person

LIBER AMICORUM PRESENTED TO

JAN HENDRIK OORT

ON THE OCCASION OF HIS 80TH BIRTHDAY

28 APRIL 1980

Edited by

HUGO VAN WOERDEN

Kapteyn Institute, Groningen

WILLEM N. BROUW

Netherlands Foundation for Radio Astronomy, Dwingeloo

and

HENK C. VAN DE HULST

Leiden Observatory

D. REIDEL PUBLISHING COMPANY

DORDRECHT : HOLLAND / BOSTON : U.S.A.

LONDON : ENGLAND

Library of Congress Cataloging in Publication Data

CIP

Main entry under title:

Oort and the universe.

Bibliography of Jan Hendrik Oort, 1922–1979: p.
Includes index.
1. Astronomy–Addresses, essays, lectures. 2. Oort, Jan Hendrik–Addresses, essays, lectures. 3. Astronomers–Netherlands–Biography. I. Oort, Jan Hendrik. II. Woerden, Hugo van. III. Brouw, Willem N., 1940– IV. Hulst, Hendrik Christoffel van de.
QB51.057 520 80–20746
ISBN 90-277-1180-1
ISBN 90-277-1209-3 (pbk.)

Published by D. Reidel Publishing Company
P.O. Box 17, 3300 AA Dordrecht, Holland

Sold and distributed in the U.S.A. and Canada
by Kluwer Boston Inc., Lincoln Building,
190 Old Derby Street, Hingham, MA 02043, U.S.A.

In all other countries, sold and distributed
by Kluwer Academic Publishers Group,
P.O. Box 322, 3300 AH Dordrecht, Holland

D. Reidel Publishing Company is a member of the Kluwer Group

Printed in The Netherlands

TABLE OF CONTENTS

TABLE OF CONTENTS

FOREWORD

Jan Oort is 80, and he is as active and creative as ever! Six months ago, he dashed into my office, with sparkling eyes and excited voice: "Come look at Ulrich Schwarz's latest results! His maps of cloud A are full of turbulence; these observations contain the key to the problem of the high-velocity clouds...". That evening he sat for hours, thinking about these new findings, then writing a proposal asking for 4x12 hours of Westerbork 21-cm line observations to study several more of these enigmatic HVC's.

A few weeks later, several colleagues met to discuss plans for celebrating Jan's eightieth birtday. We concluded that, in view of his passion for,and continuing activity in astronomy, a scientific colloquium - though modest in duration - should be an integral part of his birthday programme. We further found that we should attempt to compose a book and present it to him on that occasion. Feasts and Symposia have been organized in his name, books have been dedicated, but no book had yet been _written for_ him.

This book, then, is for Jan Oort, on his 80th birthday. It is about him and his science, about the way he works on his subject and with people. It is born of our gratitude, admiration and friendship for him, our _primus inter pares_.

We asked a selection of colleagues and friends, former students and present collaborators, to contribute to this book. We sent them a tentative table of contents, indicating suggested subjects; however, each author was left free as to title, subject matter, character and style of his contribution. We aimed at a bundle of essays of varied character, some of a personal nature, others scientific, but together illuminating many facets of Oort as a scientist.

Time was short, but the response to our invitation was enthusiastic. And here is the result. Blaauw tells the story of Oort's work, pointing out the highlights of fifty years of research. Fifty only, for Van der Laan writes about Oort's first ten post-retirement years, ten years full of astronomy. Van Bueren sets the national scene for Oort's activity (or rather: Oort sets the scene?); Strömgren and Sadler discuss his role in astronomy world-wide.

Then follow essays describing successive phases of Oort's work. Van de Kamp tells of Oort's predoctoral years in the United States. Bok and Per Olof Lindblad give close-up accounts of the discovery of galactic rotation by Oort and Bertil Lindblad. Muller recalls the beginnings of Dutch radio astronomy; Christiansen recounts Oort's struggle to obtain good angular resolution at radio wavelengths, resulting in the Westerbork Synthesis Radio Telescope. Allen and Ekers, in a profusely illustrated paper, review the highlights of Westerbork's first ten years

H. van Woerden, W. N. Brouw, and H. C. van de Hulst (eds.), Oort and the Universe, vii–viii.

- work by dozens of astronomers, with Oort's telescope and often inspired by discussions with him.

The next five contributions (by Schmidt, Woltjer, Burton, Ginzburg and the Burbidges) reflect some of the great variety of Oort's research subjects and interests: comets, supernova remnants, galactic structure, cosmic rays, external galaxies. Each of these essays represents a significant review paper.

The remaining five essays and letters elucidate aspects of Oort's personality. His son, Bram, compares the scientific characters of father and son. Bannier, former Science Administrator, shows how Oort managed to get the funds he wanted. Walraven tells the story of Oort at the telescope, Westerhout that of Oort under the skies and on skates. Van de Hulst analyzes scientists in general, and projects Oort against this background.

In these paragraphs, I have schematized the contents of this book. But Jan Oort cannot be schematized. Van Bueren, Strömgren and Bannier discuss Oort's subjects and ideas as much as the other authors. And traits of Oort's person are to be traced throughout this bundle of essays. Thus, there is considerable overlap between various contributions. We had anticipated and indeed hoped for this; for we believe we can only jointly paint a proper portrait.

Since the contributions vary greatly in character, some give bibliographies, while others don't; and we have avoided spoiling style by formalizing references to literature listed at the end of a paper. The book closes with a full bibliography of Oort's publications.
We wrote this book for Jan Oort, but also for the international astronomical community, of which Oort is and has been such an active member for now almost sixty years. We hope it will also be read by many outside this community, by physicists and other scientists, by amateur astronomers and by historians of science. Since we aimed at a wide readership, we chose for offset production from camera-ready copy.
May many enjoy the contents!

Our thanks go to Reidel for their enthusiastic support of our plan; to the authors for their prompt cooperation; to Harmen Meijer for his photographic work; and especially to Jeanet Millenaar of the Netherlands Foundation for Radio Astronomy for her tireless typing. Also, I wish to thank Wim Brouw for his great share in the editorial work; without him, I could not have done.

Finally, dear Professor Oort, beste Jan: many happy returns of the day!

Groningen, 24 april 1980 Hugo van Woerden

JAN H. OORT'S WORK
As told to the youngsters at Leiden - and elsewhere

Adriaan Blaauw

28 April 1980 - Jan Oort's 80th birthday!

We, his friends and colleagues, will celebrate it by presenting this Liber Amicorum at the conclusion of our afternoon colloquium. A special colloquium, without any doubt, but one very mich like the regular Thursday 4 o'clock Leiden Observatory colloquia. I anticipate seeing Jan Oort, in his favourite seat in the front row, listening attentively, occasionally making notes in his miniature handwriting in a mini-notebook, and intervening once in a while with a pointed remark... the way the older ones among us have witnessed him doing for some fifty years, and the youngest ones for a year or less. Sometime, somewhere his remarks will have their effect on the speaker's further work - and the notes on Oort's own thinking and research.

Jan Oort at our colloquia, that means almost sixty years of learning and research among younger colleagues and "advanced students". But what do you youngsters really know about the work that brought world fame and innumerable honours to this man? I am afraid far too little, so let me take this opportunity to present a sketch - albeit a very personally flavoured one! We shall stride with seven-league boots through half a century of astronomy, stepping from hill-top to hill-top, forgetting about completeness and details. In doing so, let us not forget that these last sixty years have encompassed almost the whole of modern astronomy. I shall dwell on the earliest decades longer than on the later ones, not only because these are more remote from the present time, but also because these early contributions in my feeling are even more fundamental than those of recent years. And as to the term "youngsters" - of course, as seen from that venerable age of 80, it will include anybody under 50.

Before we start and look at the substance of Oort's work, can we give a general characteristic of his approach to scientific research? Elsewhere in this volume, Henk van de Hulst (and others) will have something to say about this. Let me just add here that I remember how, in 1956, at the occasion of the 70th birthday of P.J. van Rhijn, Oort said that, as compared to Van Rhijn who worked so systematically along the lines programmed already in the time of Kapteyn, he considered himself rather an adventurer in galactic research. This may seem surprising to those who know Oort as the one who, probably more than anybody else,

H. van Woerden, W. N. Brouw, and H. C. van de Hulst (eds.), Oort and the Universe, 1–19.

shaped our present knowledge of the structure and contents of the Galaxy - and did so in a life-time programme. I believe that Oort's remark was prompted by his outstanding alertness to new and unexpected developments in astronomy, particularly those in the field of observational techniques that might suddenly open up new avenues to research, and that cannot be foreseen in even the best kind of "planning". His approach is sometimes intuitive and seemingly ad hoc. The mathematically oriented cosmologist, Otto Heckmann, a close friend of Oort's, once said to me "I do not understand Oort's approach, but I admire him greatly".

Let us then now go back to the early 1920's, the time when Oort entered research as a student of Kapteyn, and remember that, on several occasions, Oort has told us that his choosing astronomy rather than physics was largely due to Kapteyn's inspiring influence. No wonder that his field of research, the Galaxy, was the one to which also Kapteyn had devoted his efforts. Kapteyn's work had come to a provisional solution with the publication by Kapteyn and Van Rhijn of a model of the Galaxy. For the immediate solar neighbourhood, this was a good approximation for the thickness and density of the system as we know it now, but it still placed the centre of the system close to the Sun (and in the direction of Cygnus). In subsequent years, the present picture, with the galactic centre in the direction of Sagittarius at some 10000 pc, gradually replaced Kapteyn's - largely through the work of Harlow Shapley on the space distribution of globular clusters. As Oort wrote in 1972 in a rare historical account (a domain of our science he is not particularly keen to spend his time on, as historians have regretfully noted ...), at the time he started astronomical research, the relation of the "Kapteyn system" to Shapley's system of globular clusters was still unclear. The concept of galactic rotation had hardly been developed.

Oort's first study, in 1922, was concerned with the asymmetry in the velocity distribution of the stars having velocities with respect to the local standard of rest exceeding 65 km/sec, a phenomenon discovered some years earlier. It became the subject of Oort's thesis, defended in 1926 at Groningen with Van Rhijn as promotor. When, now four years ago, we had a small celebration at Leiden on the occasion of the 50th anniversary of his promotion, Oort remarked that he, frankly, had some misgivings about his thesis work; the reason being that the thesis did not provide the correct explanation of the phenomenon studied.

The real insight into the phenomenon of the "high-velocity stars" was prompted by Bertil Lindblad's papers. Lindblad, inspired by Shapley's work, developed a theory describing the Galaxy as consisting of a series of concentric subsystems with different degrees of flattening and different velocities of rotation; the flatter the subsystem, the faster the rotation, a concept that we encounter again in a more modern version in Baade's population-types hypothesis proposed twenty years after Lindblad's initial work. (And that, in turn, would become more refined at Oort's proposal at the Vatican Conference on Stellar Populations in 1957.)

Against these developments we must see Oort's famous subsequent papers dealing with galactic rotation. Of course, for a proper evaluation one should read these papers oneself. Note how the extensive introductory remarks help us now, fifty years later, to place them in their historical context. First, there was the 1927 paper (BAN No. 120) under the longitude, and minima at 100° and 280°, with a semi-amplitude of $\frac{3}{4}\frac{r}{R}V$. Now the most distant objects observed for radial velocity are at distances of about 1000 parsecs; with $R = 10\,000$ parsecs and $V = 300$ *km/sec* this gives a semi-amplitude of over 20 *km/sec*, which might well be verifiable. With the same assumptions the maximum effect in the proper motions in galactic longitude would be equal to $-0.''005$ per annum. The maximum will occur 90° from the direction towards the centre. In the direction of the centre and in the opposite direction the average proper motion in longitude should be equal to about $+0.''002$. The proper motion effects are, of course, independent of the distance of the objects considered.

3. *Discussion of the radial velocities.*

Several astronomers have remarked upon instances in which the stars in different parts of the sky appeared to move differently.

The hypothesis of a rotation around a distant centre has also been put forward by STRÖMBERG *) on the basis of an investigation of the preferential motions of the stars. He found that the maximum peculiar radial velocity did not occur in two exactly opposite points of the sky but in directions inclined to each other. In explanation of this he suggested a rotation around a centre near 256° longitude. Later on it has become evident, however, that these results were caused by the influence of the stars of high velocity.

In a paper on the distribution of stellar velocities GYLLENBERG pointed out that the so-called *K*-term in the radial velocities of the *B* type stars depended upon the galactic longitude **). From his drawing it is apparent that the *K*-term has distinct maxima somewhere around 0° and 180° longitude and minima at 90° and 270°. It is evident from the foregoing that this variation can be explained as the effect of rotation around a centre in 325° longitude, for the longitudes of the maxima are very near those expected in the case of rotation.

In 1922 FREUNDLICH and VON DER PAHLEN ***) have extended GYLLENBERG's investigation. According to their statements they do not doubt the reality of the variation of the *K*-term with galactic longitude, and they propose several dynamical explanations; but none of these was considered to be very satisfactory.

It is not only the velocities of the *B* stars which have given evidence of systematic motions. In a statistical study of the *c*-stars SCHILT has remarked upon the deviations from zero of the mean peculiar velocities of stars in different longitudes *). His table of residuals is reproduced below.

TABLE 1.

Average longitude	Average peculiar velocity	mean error	sin 2 $(l - 325°)$
30°	+ 8 *km/sec*	± 3.5 *km/sec*	+ .77
90	− 8	± 2.7	− .94
150	0	± 3.6	+ .17
210	+ 10	± 3.9	+ .77
270	− 7	± 4.3	− .94
330	0	± 3.5	+ .17

In the last column I have added the coefficient of the rotation term for which we are looking; it varies in very nearly the same manner as the average residual velocity.

A somewhat analogous variation has been found by HENROTEAU in a recent paper on pseudo-cepheids **).

The *O*-type stars have also been under suspicion of giving different systematic motions in different parts of the sky ***), and in this case too the general character of the residuals is what we must expect if the system of stars is rotating in the way described.

In the following table are summarized the results of a re-discussion of the radial velocities of all objects of which it might be hoped that they would show the effects of the rotation, if it exists. The second column gives the average apparent magnitude, for the *Md* variables the average maximum magnitude; in the case of the planetary nebulae it gives the limits of the apparent magnitude of the central stars. The third column shows the number of stars used, the fourth their average parallax and its mean error. Excepting the *Md* variables for which the mean parallaxes were estimated directly from R. E. WILSON's results ****), all the parallaxes were computed anew from all proper motion data avaible, in such a way as to be uninfluenced by possible rotation terms in the proper motions (see section 4). The fifth column shows the semi-amplitude of the rotational term and its mean error. In general the stars were divided into intervals of 15° or 30° galactic longitude and stars of higher galactic latitudes were excluded as mentioned in the remarks. If the longitude of the centre of a

*) *Astrophysical Journal*, 47, 32–34, 1918; *Mt Wilson Contr.* N°. 144.
**) *Lund Meddelanden* Ser. II N°. 13, pp. 22–26, 1915.
***) *Astr. Nachr.*, 218, 369–400.

*) *Bull. Astr. Inst. Netherlands*, Vol. 2, p. 50.
**) *Journal R.A.S. Canada*, 21, 1, 1927.
***) *Groningen Publications* N°. 40, pp. 52–53.
****) *Astronomical Journal*, 35, 129, 1923.

Sample page from B.A.N. 120, discussing the differential galactic rotation in radial velocities.

title "Observational evidence confirming Lindblad's hypothesis of a rotation of the galactic system", part of which is reproduced here. As it says in the introductory section: *The following paper is an attempt to verify in a direct way the fundamental hypothesis underlying Lindblad's theory, namely that of the rotation of the galactic system around a point near the centre of the system of globular clusters.* The paper demonstrates that the rotation of the Galaxy is revealed by the observed differential effects in the motions of the nearby stars, that is those within a few thousand parsecs. It shows that this interpretation agrees with the hypothesis of the centre of the Galaxy being in the direction of the centre of the globular clusters, and the paper introduces the so-called constants A and B of differential rotation - now commonly referred to as Oort's constants - which allow a measure of the local angular velocity and of the local gradient of the force directed towards the galactic centre. An important result, following immediately from the recognition of the differential rotation, is that the Galaxy is not a homogeneous spheroid rotating with constant angular velocity, for in that case there would be no differential rotation; the larger angular velocity nearer to the centre revealed the higher density in those parts. Actually, as is well known, radial velocities only give information about the constant A. The necessary additional information, embodied in the constant B, can be found only from the effect in the proper motions. The information on proper motions at that time was rather limited. Oort used it to describe the force towards the galactic centre as the sum of a part K_1, inversely proportional to the square of the distance from the centre (the Newtonian field of a point source) and another part K_2, proportional to that distance (as in a homogeneous ellipsoid). A more complete discussion of the proper motions and of the radial velocities of special categories of stars were given in Oort's subsequent papers.

The next major step is published a year and a half later, in 1928, under the title "Dynamics of the galactic system in the vicinity of the sun", BAN No. 159. From the introductory summary we quote:

"Various consequences are discussed of the theory of a rotating stellar system, the stars in our neighbourhood being considered as part of just an arbitrary section from the great system. In broad lines the discussion leads to the same conclusions as were previously reached by LINDBLAD, viz. that the star streaming as well as the systematic motions of the stars of high velocity can be considered as steady phenomena; both are to be expected a priori in such a system.

....

The stars are all considered to form part of one rotating system, but with varying peculiar velocities.

....

The view is put forward that the observed absolute limit of the high-velocity stars (at about 63 km/sec) may be identical with the difference between the velocity of escape and the circular motion in the galactic system; this is illustrated in figure 2."

Here, then, we see the breakthrough leading to the better understanding of the thesis subject of three years earlier, presented in the context of an extensive treatment of the relation between the various parameters describing the kinematical properties of the Galaxy and the

FIGURE 2.

Relation of the distribution of high velocities to the rotation of the galactic system.

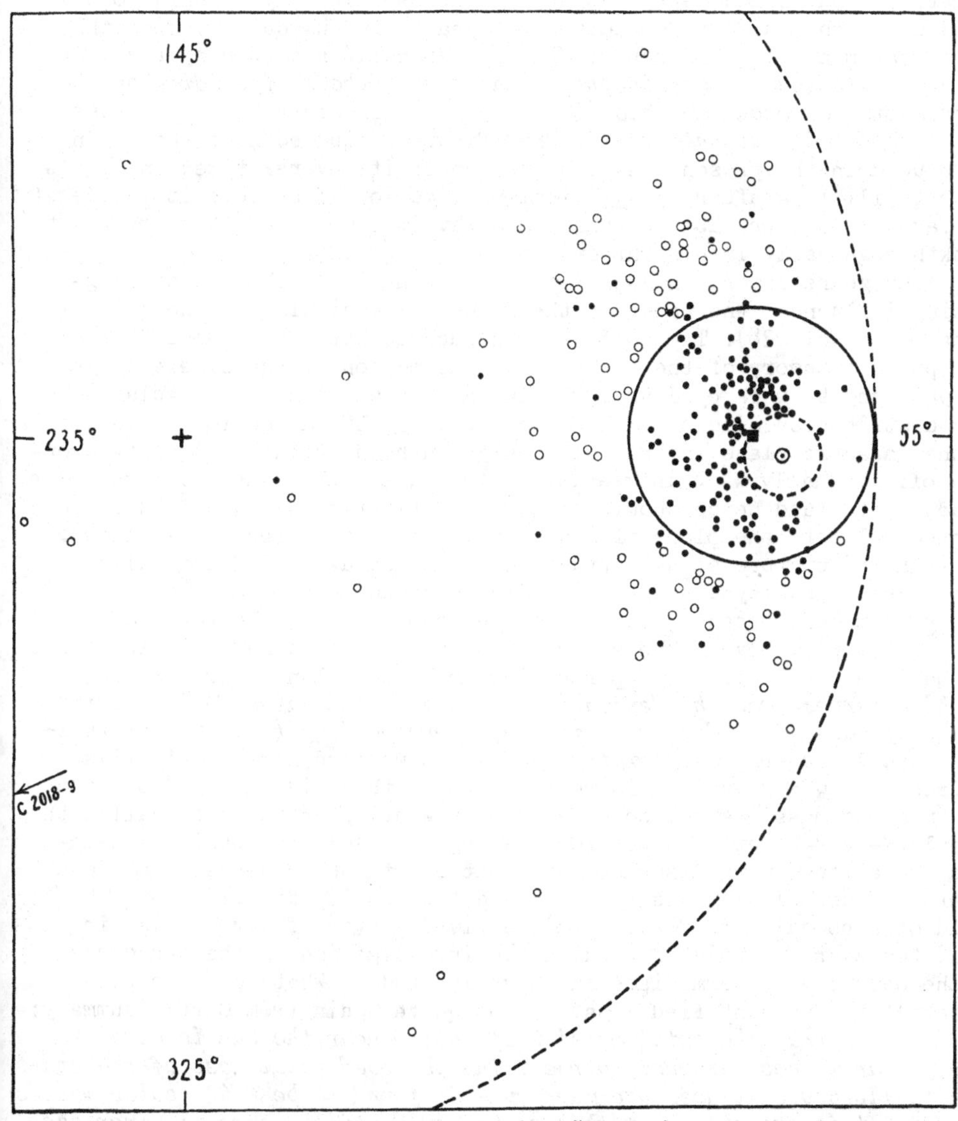

Abscissae are components of the velocities parallel to the rotation, ordinates are the components parallel to the direction of the centre. The cross represents the origin, which corresponds with the velocity of the centre of the galaxy. The distance between this cross and the black square represents the velocity of rotation for the bright stars. The full drawn circle with the square as centre has a radius corresponding to 65 km/sec and contains practically all velocities of bright stars.

The dots and small circles outside this circle show the motions of all high velocity stars for which reliable space velocities are known. The dots are supposed to represent a homogeneous-sample, complete down to 19.5 km/sec from the solar velocity (indicated by a small circle with a dot). The large broken circle has a radius of 365 km/sec and is supposed to indicate the limit beyond which velocities would surpass the velocity of escape.

The plane of drawing co-incides with the galactic plane; the numbers near the edges indicate galactic longitudes.

Key figure from B.A.N. 159

properties of its field of force, on the basis of a steady-state theory. The treatment was inspired by earlier work of Eddington, Jeans, Kapteyn and Lindblad. It included a careful confrontation with the available observational data, and represented a synthesis which may be said to have rounded off the subject for quite some years. Let us quote one of the other statements in this paper: *"The differential rotation of the galactic system is seen to be directly tied up with the ellipsoidity of the distribution of peculiar velocities. In a steady-state system the one cannot exist without the other"*. The "figure 2" quoted above has been reproduced in the present text. We encounter it several times in Oort's papers, till long after the original publication. I believe it is one of his favourites - probably because in a way it symbolizes that important breakthrough early in his career?

Four years later, in 1932, appears another contribution of long-lasting influence: the paper on the force perpendicular to the galactic plane (BAN No. 238). The subject goes back to Kapteyn's paper "First attempt at a theory of the arrangement and motion of the sidereal system" of 1922, and to work by Jeans. The basic idea was that the regular density distribution of certain types of stars in the direction perpendicular to the galactic plane in the solar neighbourhood, with the density dropping off gradually with increasing distance, suggests the existence of a steady state, and hence should allow the estimation of the field of force perpendicular to the plane if also the velocity distribution is known. Although the picture of the structure of the Galaxy had drastically changed since Kapteyn, this basic idea remained just as applicable as in Kapteyn's model, because, as was pointed out by Oort, the parts of the flat system well beyond the solar neighbourhood would have little effect on the local field of force perpendicular to the plane. Furthermore, *"A third purpose was the derivation of an accurate value for the total amount of mass, including dark matter, corresponding to a unit of luminosity in the surroundings of the sun"*. Thus we find here Oort's first reference to what nowadays is referred to as the "hidden mass" or even the "missing mass", and also reference to what later would be called the (local) M-over-L ratio. After formulating concisely the basic mathematical relations, Oort discusses at great length the observational data regarding the velocity distribution and the density distribution. The field of perpendicular force, K(z), arrived at (see figure 3) is, in turn, the starting point for extensive investigation of the structure of the Galaxy at intermediate and high latitudes. What does the paper say about the unidentified mass? Let us quote again from Oort's summary: *It is found that the total density of matter near the sun is equal to 0.092 solar masses per cubic parsec. The observed total mass of the stars down to +13.5 visual absolute magnitude is found to be 0.038 solar masses per pc*^{3}. *It is probable that this value could still be greatly increased if we could have taken the next 5 absolute magnitudes into account, so that the total mass of meteors and nebular material is probably small in comparison with that of the stars.* ----, and from the relevant section in the article: *Extrapolating the mass of the faint stars the total mass gets dangerously near the value of 0.092 solar masses derived from K(z).* As we see, Oort did not particularly worry about a "missing" mass at that time! Later, in 1965, in the chapter on Stellar Dynamics to be mentioned

FIGURE 4.

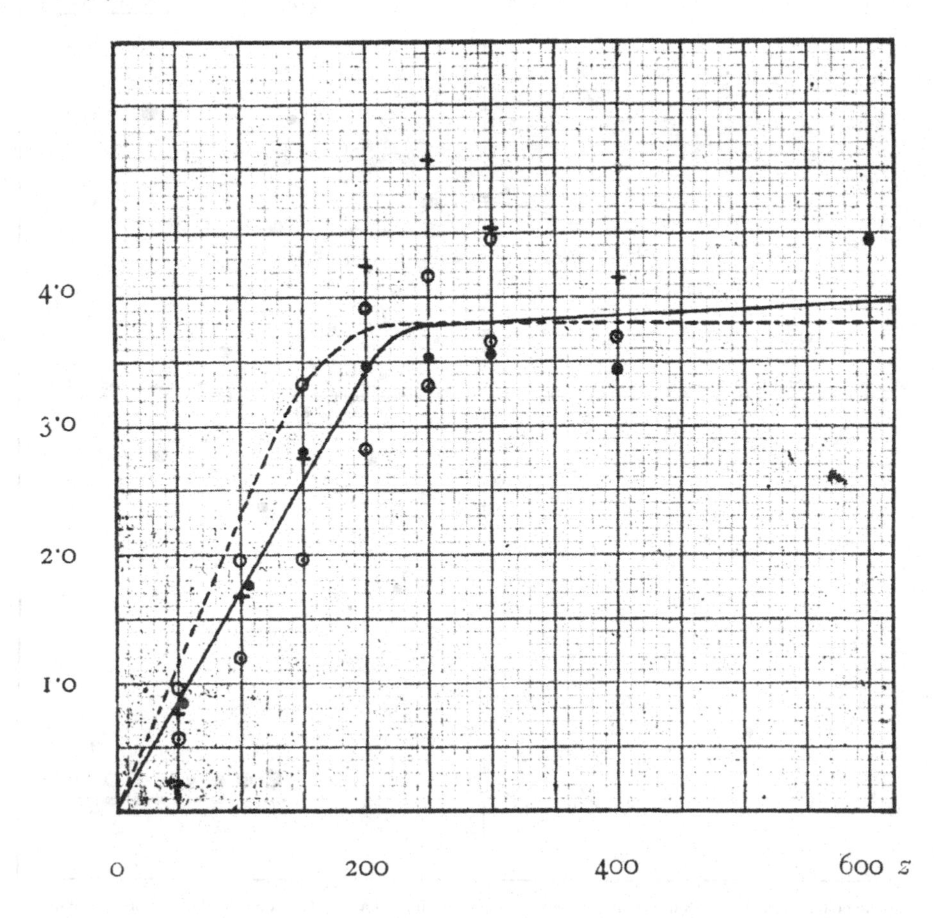

The acceleration $K(z)$.

z Is expressed in parsecs, $K(z)$ in cm/sec². The values finally adopted are represented by the full-drawn curve; the dotted curve is the preliminary result in the 2nd column of Table 14. The points plotted refer to different groups of absolute magnitude: − 1·5 to − ·5 crosses, − ·5 to + 1·5 dots, + 1·5 to + 3·5 dots surrounded by circles and + 3·5 to 5·5 circles.

Key figure from B.A.N. 238.

FIGURE 7.

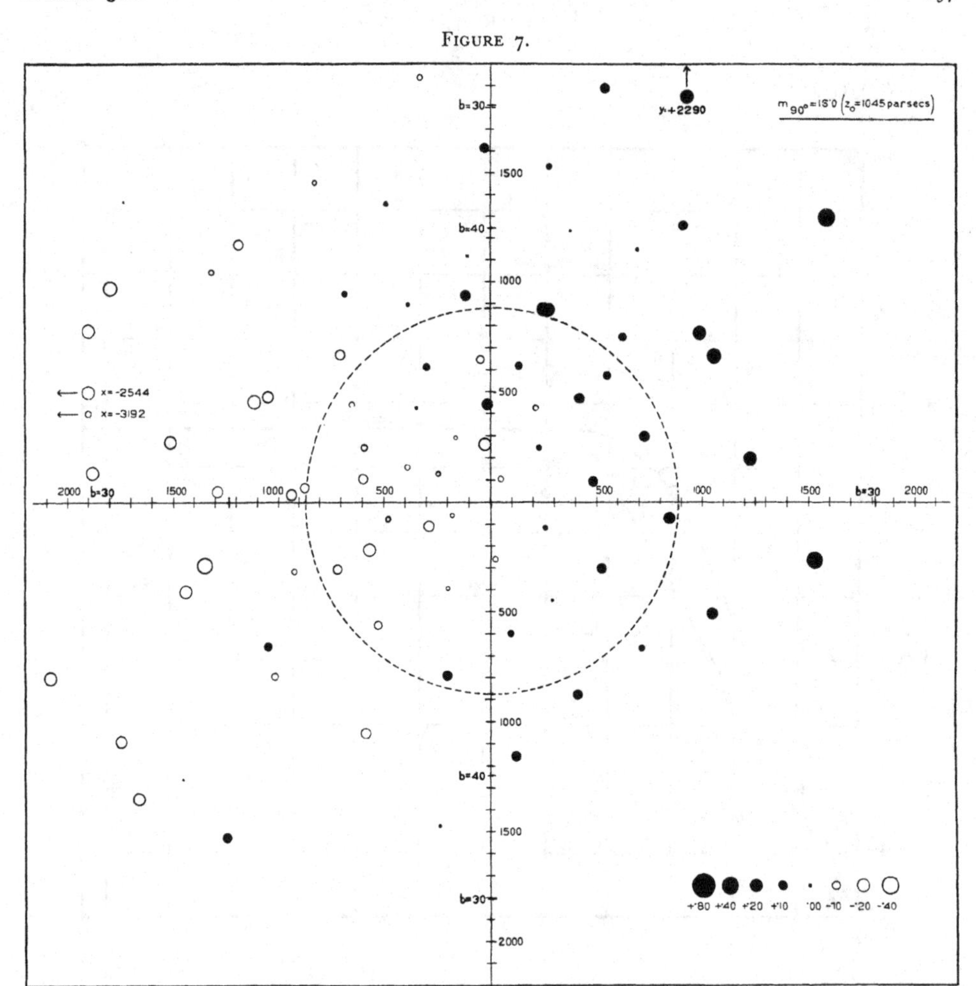

Deviations of the logarithms of the actual numbers of stars from the values corresponding to the case of equidensity surfaces parallel to the galactic plane.

Each figure shows the deviations for levels at a distance z_0 from the galactic plane; Fig. 5 corresponds to $z_0 = \pm 173$ ps, Fig. 6 to $z_0 = \pm 574$, Fig. 7 to $z_0 = \pm 1040$, areas in levels north of the galactic plane being indicated in black, those south of this plane in red.

For each level the residual for a certain Selected Area is plotted as either a circle or a dot, the centre of which corresponds to the logarithmic mean position of the stars considered, as projected on the galactic plane. The abscissæ, x_0, are projected distances measured in the direction of the galactic centre, while the y_0-axis is in the direction of 55° longitude; these co-ordinates can be computed directly from the data given in Table 8. The scale of distances has been indicated in parsecs.

The red numbers and divisions give the galactic latitudes corresponding to various distances from the centre of each diagram; the dotted red circles correspond to $b = 50°$, and delimit, therefore, the standard region.

A filled circle indicates that the number of counted stars is greater than would correspond to a plane-parallel density distribution, while open circles indicate the opposite. The area of the circles is proportional to the size of the logarithmic residual, the scale being shown in the lower right-hand corner of Figure 7. Cases where only minimum values could be computed have been designated by special signs. Dots or circles surrounded by squares represent areas in the zone at — 30° declination, and are especially uncertain.

One of the three figures showing the variations of star density with position in the Galaxy.

below, he will wonder more seriously about it, saying: *It is the unknown population that is the principal obstacle for deriving a model of the mass distribution in the Galaxy.*

The next hill-top to which we leap is a 1938 paper, "Absorption and density distribution in the galactic system" (BAN No. 308). Since the earliest work on the Galaxy, interstellar absorption had been the big obstacle in penetrating into the parts of the Galaxy beyond a thousand parsecs or so. It had been realized already in the early days, that one way to establish the presence of obscuring matter would be to measure its effect on the colours of the stars, as the amount of obscuration should be a function of wavelength. Van Rhijn's thesis project in 1917 was, in fact, aimed at establishing this effect, but the results, based on photographic photometry, had been inconclusive. The breakthrough came with the improved accuracy of photographic colours and particularly with the development of the much more accurate photoelectric photometry. Oort first of all established that the thickness of the absorbing layer must be very small, and hence that the absorption as derived from Hubble's galaxy counts at intermediate and high latitudes did take place virtually entirely in front of the stars observed at those latitudes. This then allowed a new analysis of the available star counts in the Kapteyn Selected Areas, taking this absorption into account. The procedure applied was of a differential nature, comparing counts at a certain latitude with those near the galactic pole, and it led to a determination of the general density gradient towards the Galactic Centre for layers above the galactic plane at heights from a few hundred to over one thousand parsec: a much more satisfactory determination than had been achieved previously. The method has become known as the Oort-Vashakidze method, the latter author having worked along similar lines at about the same epoch. I remember well the interest aroused by this paper when Oort presented his results at the 1938 IAU Assembly at Stockholm.

I also remember how at the General Assembly concluding these meetings the farewell words of the new president, Sir Arthur Eddington, were full of apprehension when he spoke of "the sky heavy with clouds in international politics". Soon after, when World War II broke out in 1939, we would realize how true these words had been. In the course of the next six years, while international communication and hence also the exchange of literature gradually became more and more limited, Oort turned his attention to a variety of topics. And speaking of the IAU, remember that Oort during that long period from 1935 to 1948 took care of the secretariat (with the able assistance of Heleen Kluyver, now Mrs. Van Herk), and that in later years, from 1958 to 1961, Oort served the IAU as its President.

One of these new topics was, if I remember well (and maybe this is confidential!), the project of writing a book on galactic structure! The war years, with their many restrictions, in a way invited one to take stock of what had been accomplished, and so the thought of writing such a book came naturally. Yet, the book was never finished - and I believe never progressed much beyond the table of contents. How come? Trying to explain this gives me opportunity to say something on how Oort goes about such a "project". Remember that in all research he starts strictly from the observed phenomena. No theorizing, if it is not inspired by the desire to explain what observation has revealed (and no complicated mathematical formulae, however nicely they may appear in print, if these tend to hide the substance of the discussion). Accordingly, the "book" would start with a treatment of the observational foundations of galactic research - and one of the most classical and fundamental of these is provided by the data on stellar proper motions. With characteristic thoroughness, Oort took up the task of reviewing the situation, recognized that there was a considerable discrepancy between the two then existing systems of fundamental stellar positions (that of FK3 and that of the GC), and decided to resolve the problem before going ahead with the book. The result, a careful study of the relative merits of the ways in which the two "fundamental" systems had been compiled, was published in 1943, in BAN No. 357.

The book never took shape, but we got a worthy substitute in 1965: Oort's chapter "Stellar Dynamics" in "Galactic Structure", volume V of the series "Stars and Stellar Systems". Here, in 46 pages, and within the context of other detailed chapters on a variety of aspects of the Galaxy, we find a masterly, concise treatment of the substance of galactic dynamics. There is no better way do describe the delight of "galactic" astronomers about the appearance of this text than by quoting the reviewer of the volume, Bart Bok, in Sky and Telescope of June, 1966:

In this reviewer's opinion, Oort's review of the progress made over the past 40 years in the field of galactic dynamics is truly magnificent. It starts from scratch, and in concise form, without derivation of equations, develops all of the needed basic concepts before embarking upon interpretation. It is a wholly gravitational treatment, applicable to stars after formation and not to the interstellar gas. This chapter deserves to be read and reread by all students of stellar dynamics. .... Through it all, there breathes the intellect and understanding of complex phenomena that have made Oort one of the great scientists of our age.

Now, fifteen years after its publication, this chapter is still as much a must for every student of galactic structure (and in fact, of the structure of other stellar systems) as it was at the time of its appearance. In a way, this chapter crowned more than forty years of Oort's work in the field and, needless to say, his own contributions formed a substantial component of the chapter. Among his more recent publications related to problems of stellar dynamics is the work with G. van Herk on the structure and composition of the globular cluster Messier 3 - another of these "musts" for workers in the field. But meanwhile Oort had developed other interests and opened up new avenues of research: Interstellar Matter and Radio Astronomy.

Oort's interest in the properties of the interstellar medium was another subject that had developed during the war years. Interstellar dust changed from an obscuring nuisance to a challenge when Oort, in collaboration with physicists (Kramers, Ter Haar and others) and Van de Hulst, took up the problem of the interaction between gas and dust, and particularly that of the formation or evaporation of dust under interstellar conditions. This resulted in the paper with Van de Hulst in 1946 (BAN No. 376). A fine opportunity to expose his thoughts about the interstellar medium offered itself when Oort was invited to deliver the George

Darwin Lecture in 1946. It appeared under the title "Some Phenomena connected with Interstellar Matter", and contains many elements which would later, either by himself or by others, be taken up for further research. It is another of those texts still worth rereading!

Soon after World War II had ended, radio astronomy assumed the important position in astronomy that it has now - after having gone through its earliest developments almost unnoticed by astronomers. When Grote Reber's article on Cosmic Static appeared in the Astrophysical Journal in 1940, Oort was among the few to immediately notice its great significance to astronomy. The story is well known: Oort realized how important it would be in a galactic context to detect interstellar radio <u>line</u> emission, as the observed wavelength would indicate not only the velocity of the source, but also its position in the Galaxy. At Oort's request, Van de Hulst considered what spectral lines might be observable at radio wavelengths, and this led to the prediction in 1944 of the 21-cm HI line. The further pursuit of this new avenue would drastically change astronomy in the Netherlands.

The change meant more than the generation of astronomers that has grown up in Holland in the post-war years may be aware of. It was not just the addition of many octaves to the wave-length domain accessible to the astronomer - it also turned astronomy in Holland into an observational science, the Dutch sites being as good as any place on Earth to measure celestial radio radiation. In the preceding decades optical astronomy in Holland had fallen more and more behind. Leiden's observational contributions had been mainly based on the 33-cm astrometric and 26-cm visual refractors and the meridian circle, all dating from the 19th century and located in a very poor climate, though in addition there was the Leiden Station in South Africa which did work under favourable climatological conditions. But astronomy as an observational science was centred on the great American observatories. Apart from purely theoretical work, astronomy in Holland had adapted itself to making use of observational data largely provided by others - be it in the form of published catalogues, or photographic plates taken by foreign institutes.

As we all know, Oort has been the principal initiator of observational radio astronomy in Holland and, when in succession the observatories at Kootwijk, Dwingeloo and Westerbork became operational, he became himself one of the principal proposers of observing programmes. Thus, Oort became an observational astronomer in the 1950's and I cannot resist the temptation to dwell for a moment on this change, for it gives me the opportunity to point out another wonderful aspect of Oort's work.

The great progress in galactic research due to Oort's contributions, as I have sketched them in the preceding paragraphs, was virtually entirely based on his analyzing, pondering about - "piekeren over" as Oort likes to say in Dutch - data that had been available in the astronomical literature for everybody, sometimes for many years. The merit in Oort's dealing with these data lies in the fact that, after learning to know them thoroughly by critical evaluation, he turns them into building blocks for newly developed insights into galactic structure and dynamics. A very good example of this is the paper on the force perpendicular to the galactic plane. In short, these papers show how much could be done by mere reflection on the data at hand; by the process of powerful thinking. It makes one wonder, how beneficial it would be for modern astronomy if the tap on the stream of new observational data would be closed now for a while, and astronomers be forced just to think about what they have already .... I am sure Jan Oort would object, for I know few people who wait so impatiently for new data as he does - but I would also trust no one better than him, just to think about what we have! And - let me also add what Jan Oort is reported to have said in a (Dutch) newspaper interview on the occasion of his retirement:

If someone adds up what I have done in half a century, then perhaps he may conclude that it is quite a lot. If I add it up myself, then I arrive at the conlusion that I might have done much more, if only I had reflected better at the right moment.

As to the lack of adequate optical facilities, Oort of course fully realized how badly we were in need of them. And so when Walter Baade, during a stay at Leiden Observatory in the spring of 1953, suggested that European astronomers should join efforts and establish a common observatory, Oort took up the suggestion with enthusiasm and called a meeting of prominent European colleagues. It led, after many years of struggle - so characteristic, unfortunately, of many efforts in European collaboration -, to the erection of the European Southern Observatory. For many years Oort has been a stimulating president of the ESO Council.

With the era of radio astronomy we enter a new phase in Oort's career. One would almost say that a new astronomer Oort emerged. The new era was marked as much by his drive and organisational talents as by his guidance in the interpretation of observations. Others (Bannier, Muller and Christiansen) will tell more about his role in setting up the radio observatories in Holland, but let me just mention one event, typical of Oort's foresight. It happened at one of the regular meetings of the Board of the Netherlands Foundation for Radio Astronomy; Oort, travelling abroad, could not be present. Under the vice-chairman's direction the meeting might have become a routine affair; the Dwingeloo telescope had recently come into regular operation, and a period of gathering the fruits seemed ahead. But Oort had left a letter, telling the Board that now the time had come for preparing for the next step, with much higher angular resolution than we had achieved so far. It led, ultimately, to the Westerbork Synthesis Telescope.

Radio astronomy would become to a large extent a joint affair with many other colleagues, Dutch and foreign, in contrast to the strongly single-handled achievements of Oort's "first phase". Fascinating results on the 21-cm emission line appeared in rapid succession. After the detection of the emission had been announced in 1951 almost simultaneously by teams in the United States (Ewen and Purcell), in Holland (Muller and Oort) and Australia (Christiansen and Hindman), the Dutch effort concentrated vigorously on the problem of the large-scale distribution of the neutral hydrogen in the Galaxy. The first observations, made with an old German radar mirror of 7.5-meter aperture at Kootwijk, resulted in two papers reflecting these efforts: "The spiral structure of the outer parts of the Galactic System ...." by Van de Hulst, Muller and Oort (BAN No. 452), and "The rotation of the inner parts of the Galactic System", by Kwee, Muller and Westerhout (BAN No. 458), both published in 1954. As it says in the introduction to the first-mentioned paper, Muller was

Oort "piekers" at his desk in 1953.

chiefly responsible for the instrumentation, Van de Hulst for the reduction of the observations, and Oort "for the astronomical discussion". That Oort also had a considerable share in the second paper is clear from its concluding paragraph.

Apart from the authors mentioned, many young astronomers, technical staff and students participated in the reduction of the observations which, at that time, was still virtually entirely "hand reduction". These were exciting years at Leiden Observatory, under the inspiring leadership of Oort and Van de Hulst, because what emerged were two new, fundamental pieces of information: the spiral structure in our Galaxy and the rotation curve for the inner parts of the system. The excitement was, in fact, world-wide, for independent of the radio work the occurrence of spiral structure in the Galaxy had in 1951 been established for the region up to 2.5 kiloparsec from the Sun by W.W. Morgan, Sharpless and Osterbrock from a study of the space distribution of OB stars.

A flow of 21-cm papers followed these first ones, including those by Westerhout on the HI distribution in the outer parts of the Galaxy, extending the work of BAN No. 452 to the hydrogen above and below the galactic plane, and by Schmidt on the spiral structure in the inner parts of the Galaxy, both in BAN No. 475 (1957). They culminated in the colour print between the papers of Westerhout and Schmidt, showing the contours of equal density as seen from the north galactic pole as far as measurable from the Netherlands. This picture was soon to be supplemented by the southern part observed in Australia. The new situation, an enormous step forward in galactic research, was then reviewed in 1958 by Oort, Kerr and Westerhout in the paper "The Galactic System as a spiral nebula" in the series Reports on Progress in Astronomy in the Monthly Notices of the R.A.S. After summarizing the new knowledge, the authors conclude the paper with a section on "Evolution of spiral structure". This structure now being established, together with the rotation curve, the time had come indeed to take up again the problem of the origin of spiral structure.

Naturally, this problem would have to be studied in the context of the universality of the phenomenon in extragalactic systems. The problem had intrigued Oort since long. In his inaugural lecture of 1935, when he took up the professorship at Leiden University, he was compelled to conclude that *the language of the spiral nebulae is still unreadable for us*. Now, in the late 1950's, the time for deciphering had arrived. The successful work of C.C. Lin and collaborators, building on ideas already explored by B. Lindblad, is hard to imagine without the galactic 21-cm results.

Apart from these major advances, the importance of which was immediately evident, the 21-cm work had revealed an other, unexpected, phenomenon that caught Oort's attention right away and was pursued by him with characteristic alertness: the remarkable features in the central region of the Galaxy. A first indication of the occurrence of strongly turbulent motions in this region had been found in the profiles obtained with the Kootwijk telescope, but after the Dwingeloo telescope with its higher resolving power came into operation in 1956, their nature could be investigated more profitably. Milestones in this work were Rougoor's

MONTHLY NOTICES OF R.A.S. VOL. 118, PLATE 6

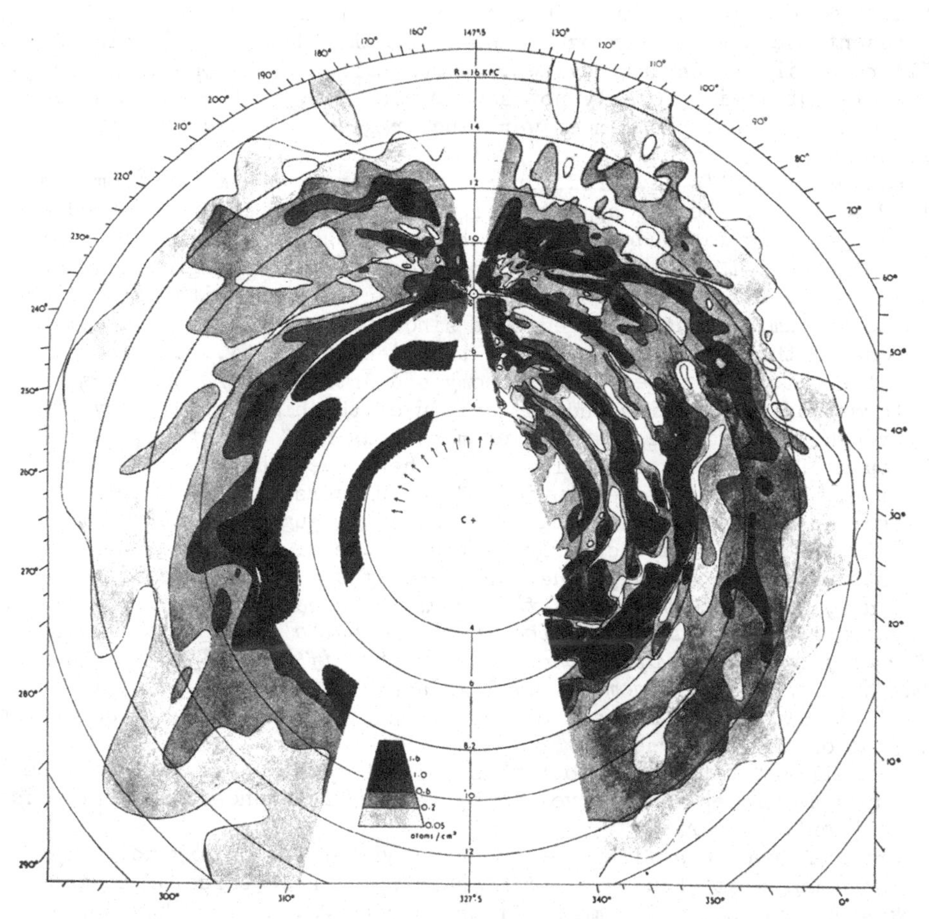

FIG. 4.—*Distribution of neutral hydrogen in the Galactic System. The maximum densities in the z-direction are projected on the galactic plane, and contours are drawn through the points.*

The first complete map of neutral hydrogen in our Galaxy (Oort, Kerr and Westerhout, 1958).

thesis of 1964 and Van der Kruit's thesis of 1971. The region within 4 kiloparsec from the Galactic Centre has become one to which Oort in various papers has given repeated and penetrating attention. The phenomena include the expanding "3 kpc" and "+135 km/sec" arms, the rotating disk within 700 pc, and various structural and kinematic features on a smaller scale, among which those revealing the ejection of matter are the most intriguing. Many of us will remember Oort's review of this subject at the occasion of the 50th anniversary of his doctorate in 1976. The present picture is thoroughly and critically described in his recent (1977) paper in the Annual Reviews. It summarizes work done on a worldwide scale but it is probably not saying too much, that quite a number of the observational programmes were encouraged by Oort's stimulating interest.

Another of Oort's favoured topics, one that also emerged from the early 21-cm work, are the high-velocity clouds observed at intermediate and high galactic latitudes. Here, Hulsbosch's thesis of 1973 was a milestone based on Dwingeloo observations since 1963. The origin of these remarkable objects, falling towards the galactic plane with velocities of up to 200 km/sec, has puzzled Oort since their discovery - and is still one of those problems he likes to "pieker" about .... even ten years after his 1970 paper in Astronomy and Astrophysics, on the subject: "The formation of galaxies and the origin of the high-velocity hydrogen".

Throughout all these years, over more than half a century, at the background of all these investigations, there has been the all-embracing problem of the origin and evolution of the Universe and the formation of galaxies. We find Oort referring to it in his inaugural lecture of 1935, and what strikes him particularly is the nature of the irregularities in the distribution on the sky: *The structure of the Universe as it is revealed by the distribution of the galaxies is hard to describe in a few sentences; it is an indescribable chaos, but a chaos that is quite different from the situation one would observe if the galaxies were distributed in space at random.* ..... This aspect has continued to intrigue him till today, and is a basic element in his current thoughts about the evolution of the Universe.

We shall not tell more about these recent researches; they will be taken up again later in this volume by Van der Laan under the appropriate heading "Meritus Emeritus".

Oort has devoted a life-time to the study of the Galaxy and the Universe - but what about spare-time hobbies? I know three of them: ice-skating, rowing, and comets. Of the first two, as he likes to point out himself, his successful performance is not unrelated to his lack of physical weight. His success in the latter we owe to the obvious presence of his intellectual weight. This hobby developed in the late 1940's and, if I remember well, started when a Leiden student interested in celestial mechanics, A.J.J. van Woerkom, wished to work on comet orbits. Oort became the one to supervise the work - perhaps not an obvious choice, but certainly a very fortunate one. It aroused his interest, presumably because of a common feature between the comet system and stellar dynamics: the influence of stellar encounters. It led to Oort's developing the theory of what is now referred to as the Oort cloud of comets, and, in collaboration with Maarten Schmidt, to the distinction between "young" and

Oort lectures on the occasion of his golden doctorate, 1 June 1976.

"old" comets.

In the interview on the occasion of Oort's retirement mentioned before, Oort is quoted to have said: .... *the piece of investigation that has given me the most satisfaction is something rather outside my regular field. It is my work on the origin of the comets in the solar system. It is the only investigation that has been properly rounded off. All other researches result in something of which, after all, one understands only half.*

Whether the last statement is true or not we shall not take up here - in any case the half of which we did gain understanding we owe to Oort in so many cases!

Among those whose signature is engraved on the astronomy of this century, Jan Oort is one of the prominent ones. That familiar signature, just four characters preceded by two initials, simple and usually without any underlining, is always clearly written, be it on a postcard with greetings from a conference, under a letter telling about his ideas or criticisms, or under a weighty document. It symbolizes the clarity of his insight and of the style of presentation with which he has let us share in the fruits of his many talents. We hope it may mark many more messages in the coming years - but above all we wish for many more years of Oort's presence among us, colleagues and students at Leiden Observatory.

References

Bok, B.J. 1966, Sky and Telescope 31, pp. 357.
Christiansen, W.N. and Hindman, J.V. 1951, Nature 168, pp. 358.
Ewen, H.I. and Purcell, E.M. 1951, Nature 168, pp. 356.
Hulsbosch, A.N.M. 1973, "Studies on high-velocity clouds" (Thesis, Leiden University).
Hulsbosch, A.N.M. 1975, Astron. Astrophys. 40, pp. 1.
Kapteyn, J.C. 1922, Astrophys. J. 55, pp. 302.
Kwee, K.K., Muller, C.A. and Westerhout, G. 1954, Bull. Astr. Inst. Netherlands 12, pp. 211 (No. 458).
Morgan, W.W., Sharpless, S. and Osterbrock, D. 1952, Astron. J. 57, pp. 3.
Muller, C.A. and Oort, J.H. 1951, Nature 168, pp. 357.
Oort, J.H. 1922, Bull. Astr. Inst. Netherlands 1, pp. 133 (No. 23).
Oort, J.H. 1926, "The stars of high velocity" (Thesis, Groningen University), Publ. Kapteyn Astr. Lab. Groningen, No. 40.
Oort, J.H. 1927, Bull. Astr. Inst. Netherlands 3, pp. 275 (No. 120).
Oort, J.H. 1928, Bull. Astr. Inst. Netherlands 4, pp. 269 (No. 159).
Oort, J.H. 1932, Bull. Astr. Inst. Netherlands 6, pp. 249 (No. 238).
Oort, J.H. 1935, "De bouw der sterrenstelsels" (Inaugural Lecture, Leiden University); Hemel en Dampkring 34, pp. 16, 1936.
Oort, J.H. 1938, Bull. Astr. Inst. Netherlands 8, pp. 233 (No. 308).
Oort, J.H. 1943, Bull. Astr. Inst. Netherlands 9, pp. 424 (No. 357).
Oort, J.H. 1946, Mon. Not. Roy. Astr. Soc. 106, pp. 159.
Oort, J.H. 1958, in: "Stellar Populations", Ed. D.J.K. O'Connell, North Holland Publ., Amsterdam, pp. 507.
Oort, J.H. 1965, in: "Galactic Structure", Eds. A. Blaauw and M. Schmidt (Stars and Stellar Systems, Vol. 5), Univ. Chicago Press, pp. 455.
Oort, J.H. 1970, Astron. Astrophys. 7, pp. 381.
Oort, J.H. 1970, Interview with "Nieuwsblad van het Noorden", Groningen, 19 September 1970.
Oort, J.H. 1972, Ann. New York Acad. Sci. 198, pp. 255.
Oort, J.H. 1977, Ann. Rev. Astron. Astrophys. 15, pp. 295.
Oort, J.H. and van Herk, G. 1959, Bull. Astr. Inst. Netherlands 14, pp. 299 (No. 491).
Oort, J.H. and van de Hulst, H.C. 1946, Bull. Astr. Inst. Netherlands 10, pp. 187 (No. 376).
Oort, J.H., Kerr, F.J. and Westerhout, G. 1958, Mon. Not. Roy. Astr. Soc. 118, pp. 379.
Reber, G. 1940, Astrophys. J. 91, pp. 621.
Rougoor, G.W. 1964, Bull. Astr. Inst. Netherlands 17, pp. 381.
Schmidt, M. 1957, Bull. Astr. Inst. Netherlands 13, pp. 247 (No. 475).
Van de Hulst, H.C. 1945, Ned. Tijdschr. Natuurk. 11, pp. 210.
Van de Hulst, H.C., Muller, C.A. and Oort, J.H. 1954, Bull. Astr. Inst. Netherlands 12, pp. 117 (No. 452).
Van der Kruit, P.C. 1970, Astron. Astrophys. 4, pp. 462.
Van der Kruit, P.C. 1971, Astron. Astrophys. 13, pp. 405.
Van Rhijn, P.J. 1917, Publ. Kapteyn Astr. Lab. Groningen, No. 27.
Westerhout, G. 1957, Bull. Astr. Inst. Netherlands 13, pp. 201 (No. 475).

Adriaan Blaauw studied at Leiden (1932-38) and obtained his doctorate at Groningen (1946). He worked at Leiden (1945-53), Yerkes and MacDonald (1953-57) and Groningen (1957-70), then was Director General of the European Southern Observatory (1970-75). He is now Professor of Astronomy at Leiden.

Adriaan Blaauw

MERITUS EMERITUS

The first decade of Jan Oort's retirement, a personal account by

Harry van der Laan

Ladies and Gentlemen, assistants and students in astronomy.... All of us have their own limited problems and therein our own difficulties. Are we to make something sound of our labours, then this will demand nearly all of our resources. Nearly, *but not entirely. I hope that we shall always, in addition, maintain the opportunity of learning eachother's problems and keep alive the awareness that all of us together are building towards one great work*

J.H.Oort, Leiden
Inaugural Lecture, 1935.

Jan Hendrik Oort, professor of astronomy and director of Leiden Observatory, retired from office as of 1 September 1970. A great many duties were lifted from him: a burden of committee-work, the worries about fifty employees and as many students, the leadership of the Dwingeloo/Westerbork radio observatories and laboratories, the remote control of the Leiden Southern Station at Hartebeespoortdam near Pretoria among them. What a relief, what an opportunity to do in full measure what comes naturally: take all this time and energy, released from inevitable and oft onerous duties, and devote them to .....research in astronomy.

The academic year 1969-70 was exciting and exhausting. In the Netherlands Foundation for Radio Astronomy a decade of planning, designing and building drew to a close as construction of the Synthesis Radio Telescope at Westerbork neared completion. At that time major changes in organization of the NFRA team were called for by younger members of staff and Board. Oort, as first and only chairman, was responsive to these pressing needs and guided the Board through a process of difficult reorganization decisions. Exceptional are people, at any age, with the flexibility and vision to choose the difficult way of change over the temptations of the status quo. On 24 June 1970 Oort was the charming host to Queen Juliana, who dedicated the Westerbork telescope. A new exploration was on its way and Oort was there to share the adventure, all the more fully as he resigned his chairman's post

H. van Woerden, W. N. Brouw, and H. C. van de Hulst (eds.), Oort and the Universe, 21–29.

six months after telescope dedication.

In April 1970 an all too rare *semaine d'étude* in astronomy took place in the Vatican Academy. The theme, chosen by Oort, was *active nuclei*, interpreted widely enough to include all, from the Galactic Centre to quasars. For an unforgettable week two dozen of us worked in the fresco-covered conference room of the academy building, itself a small architectural jewel set in the breathtaking papal gardens whose stillness is enhanced by the immense dome of St. Peter's. Long working days, eased by sumptuous coffee breaks and touristic lunches, resulted in a thousand-page book that continues its instructive rôle despite the turbulent advance of another decade.

Following this intensive Roman week, with only a weekend at Leiden for sorting mail and things, we all went to Groningen. There, about seventy friends celebrated Oort's seventieth birthday at a symposium in Roden. A highlight of this meeting was Oort's lecture, in the imposing hall of Groningen's Academy, entitled *Galaxies and the Universe*. I will return to this lecture later in this chapter, for it is a revealing tour, much more the description of a programme than statement of departure at the close of a successful career.

More celebrations were held in the course of that academic year, including an oldfashioned *Sterrewacht-feest*. All the while there were daily duties and the prosaic activities of pushing contractors to finish the new house and preparing the move. After decades of family life in the director's residence at the Sterrewacht the Oorts had decided to build a house on Oegstgeest's Kennedylaan, with a generous garden along the canal. The observatory's cramped quarters were to be extended by conversion of the director's house into office space. The timing of housebuilding in Oegstgeest and painting at the Sterrewacht went awry, and the Oorts moved into their home while plumbers and carpenters were still making promises. It was all very exhausting and friends watched worriedly: after all, seventy is no age to build and move house, to write and to lecture, to travel and throw parties all in a short span of time. Or is it?

The unbroken sequence of Oort's IAU General Assemblies continued in Brighton that summer and then, in September, life assumed a new routine. Settling in on the Kennedylaan Mieke Oort began another decade of cheerful hospitality for numerous guests, a spectrum from prominent foreign friends to first-year Leiden research students. Jan Oort now took a fifteen-minute early-morning bike ride instead of the thirty strides that so long separated home from observatory office.

It took time, I think for all of us, to be comfortable in the new pattern of things. Oort had cleared the imposing room with which so many of us associated him: booklined walls, a large desk and a conference table neatly piled with sorted papers, letters and notebooks. A spacious corner with oldfashioned lounge chairs for quiet conversation, a bouquet of flowers or budding branches in large vases. The room was usually in semidarkness, overgrown awnings filtering the light from oft cloudy skies. De Sitter's stern features overlooked the whole room from the lit painting above the mantlepiece. In that room the brightest spot was from a large desk lamp illuminating a small oak writingtable. At that table current research was done, drafts were written; speculations,

Oort leaving the old building of the Sterrewacht te Leiden.

hunches became observing proposals, intuition spawned insights. I remember how awed I was by this room and its occupant; after many North-American years I sensed here the quintessence of European erudition.

His new office was a brightly painted cornerroom upstairs in their former house, with a fraction of his old furniture and new grey-metal bookcase. I was troubled; if Oort was, he did not say so.

Leiden astronomy began a new period of excitement those days. The Westerbork telescope observed full time and maps poured forth from the university computing centre. Radio galaxies we had long known from Cambridge showed new lobes or bridges; quasars were subtly structured triples, galaxies in clusters left long twisted trails and in many a 'normal' galaxy we found an active nucleus. In the Milky Way, supernova shells in highly polarized patterns suggested how their dynamics went, and radio blisters on H II regions led to systematic surveys, preludes to studies in starformation. At coffeetime in the former large meridian-circle room, surprise after surprise was spread on the ping-pong table. In all these things Oort took a keen interest, participating in lunch discussions, informal colloquia and all-day sessions, nowadays called workshops. This interest and involvement did not, for these subjects, result in his joining the analysis, in authoring Westerbork-based research papers. Oort monitored our progress, stimulated and encouraged, reporting occasionally at the monthly meetings of the Royal Academy. His own work continued on Dwingeloo surveys of high-velocity clouds and on a variety of long-term programmes aimed at classical problems such as M-dwarfs in the solar neighbourhood, Hyades Cluster outer members, RR Lyrae distribution in the inner Milky Way.

Oort's Westerbork-related interests grew to active participating labour in the area of nearby spiral galaxies. The Westerbork SRT opened up this area of research, thanks to its unique combination of angular resolution and sensitivity, naturally continuing the great tradition of galactic research established in Dutch astronomy since Kapteyn. Very quickly a major survey in 1415 MHz continuum radiation was on the way and yielding beautiful new results. The nonthermal radio-continuum spiral pattern in Messier 51 became famous in a matter of weeks throughout the astronomical world, a great stimulus for theoretical work on spiral density waves. At first Oort coached, coaxed and gently pushed the workers at Leiden and Groningen to publish their results, but did not see or take an opening to join in the fray. Not until NGC 4258 burst upon the scene, with its great nonthermal radio wings emanating in rotationally symmetric fashion from the stubby optical HII arms discovered ten years earlier in France. Here finally a direct connection seemed revealed between explosive nuclei and spiral structure: NGC 4258 was instantly promoted to most beautiful phenomenon in the sky and Oort co-authored his first major Westerbork research paper. In addition he wrote reviews, constantly incorporating the latest results, the newest maps into his comprehensive grasp. The 1972 Karl Schwarzschild Vorlesung and his reviews at the Athens Regional Meeting of the IAU the same year gave flavour and substance of those heady days. Invited papers at the Canberra IAU symposium no. 58 in 1973 and at the CNRS colloquium at Bures-sur-Yvette the following year show the nuances and difficulties raised by the new radio phenomenology.

Spiral structure, its kinematics and dynamics, its origin and maintenance forms a major theme of Oort's lifelong research. It is nevertheless *not* his most basic astronomical interest, nor the problem area to look in search of the motive power of his relentless drive and seemingly tireless efforts. That area is cosmology. I shall come to that, but first must relate some matters more mundane.

In 1973-74 we struggled as an institute against our Faculty- and Central University Boards to keep our home, the beautiful Sterrewacht built on what was formerly one of the old city's fortification sites, adjacent to the *Hortus Botanicus*. The main building, completed in 1861, had been extended by a number of annexes, and workspace had grown by successive conversions of homes, the Walravens', the Oosterhoffs' and the Oorts', into offices. The building was adequate till the end of the century; the city centre, bookshops, student café's were conveniently close and only a small minority of the Sterrewacht community was prepared to go, as an afterthought, into the Huygens Laboratory outside the city, a new physics building planned in extrapolation during the booming sixties and threatening to remain onethird empty upon completion.

The easy coincidence of departmental, institutional and management boundaries, the creative atmosphere, the historic sense of pride, 'the living awareness that all of us together are building towards one great work', in short our *esprit de corps* were severely damaged in this struggle which we lost against uncomprehending technocrats and bureaucrats. They haven't the faintest idea of the cost ....Oort did not involve himself directly in this fight, except when one day he heard that upon leaving the *building* called Sterrewacht, we would also lose our *name* and henceforth be the astronomy department in a physics building. He came to me, shaking with barely contained rage: "You don't ever give up your name! Sterrewacht te Leiden, *Leiden Observatory* cannot be erased from astronomy, think of Kaiser, of Hertzsprung, of De Sitter ......" I thought of Oort, too. We kept our institutional name, licked our wounds and moved. I spent a year at the Institute in Princeton, to work quietly and to let my sense of outrage fade. It never quite did.

In the course of the 1970's the Westerbork SRT was steadily diversified and improved. From a single-frequency continuum telescope it became a three-frequency polarimeter and a high-resolution radio-spectrophotometer. Sensitivity limits were lowered time and again; plans for baseline extensions were developed and began to be implemented. VLBI became part of the repertoire. Oort had resigned from most functions in Leiden, nationally and worldwide upon official retirement. At the urging of its members he has been repeatedly persuaded to remain a member of the Programme Committee, which allocates time on the Dwingeloo and Westerbork radio telescopes. In ten years he has attended as many of its meetings as anyone of us. He clearly enjoys weighing the importance of an idea against the cost in telescope time required to confirm or refute it. He does his refereeing homework conscientiously. Once or twice by telephone from his bedside when flu or a cold, and Mieke's authority, kept him there. Our PC meetings have been greatly enhanced by his presence and lively participation. He never hides his preferences, is unabashed in the advocacy of his favourite topics and in ten years has pronounced each of more than half a dozen programmes the single most important proposal ever to reach the PC! No doubt the weighing of hundreds of observing programmes on the Dwingeloo and Westerbork telescopes has these ten years benefited greatly from his experience, his criticism, and his legendary sense for important problems and key observations.

Let us return to the quotation of the thirty-five year old Oort which began this account. He speaks there of problems and difficulties, of strenuous efforts requiring all our energy if they are to produce worthwhile results. Good research in astronomy means toil, means sacrifices of comfort, of time for family, friends, hobbies, relaxation. The Churchillian blood, sweat and tears phrase is too dramatic here, but then only because our work is so irenic. By Oort's own maxim, sound research is also hard work; he never hides the fact that he finds writing difficult. The reader now only need inspect the Oort bibliography at the end of this book to fathom the immense amount of labour done this decade. The single most sustained effort was on the Galactic Centre, an Oort favourite for many years. At the Royal Greenwich Observatory tercentenary celebration in July 1975 he gave a comprehensive, profusely illustrated lecture on this topic. He regarded this as only a preliminary trial run, a semi-popular contribution. Less than a year later he started a paper, review and research paper both, that covers nearly seventy (70!) printed pages in the 1977 Annual Review of Astronomy and Astrophysics. "Are we to make something sound of our labours, then this will demand nearly all of our resources". In 1976 Oort experienced again what he had said to astronomy students in 1935. The paper is sound, synthesizing, in what is at once a grand overview and a meticulous arrangement of detail, the manifold jumble of facts and figures about the Galactic Centre that came from all parts of the electromagnetic spectrum. The preprint, widely distributed and multiply copied, was an instant hit among the many teams working on aspects of the Galactic Centre. I know, from their own comments, that it gave many of these young people a renewed sense of what they were doing, of seeing the big puzzle while they struggled with one piece. They also gratefully accepted the chance of henceforth ignoring most Galactic Centre literature prior to 1976.

Earlier I claimed that *spiral structure* is not the theme that moves Oort most. I now add that *active galaxy nuclei*, starting with the Galactic Centre, does not as a theme play that central rôle either. They are important enough and especially in combination greatly excite Oort, witness the lavish attention he has given NGC 4258. His repeated efforts to understand NGC 1275 are especially noteworthy and his interests in all active nuclei from the Galaxy to M 81 to M 87 to remote radio galaxies are well documented in numerous reviews. Yet they are all, in a sense, preludes to his deeper interests: cosmology and the successive origin of structure in the Universe. Not that Oort has done all that much cosmology research himself. He knows as well as anyone that observational cosmology is the collective effect of many laboriously detailed programmes on carefully selected samples; a struggle against selection effects, the great spread of luminosity functions and of mass-to-luminosity ratios. His Solvay-Conference paper *Distributions of Galaxies and the Density in the Universe*, published in 1958, is the result of

Oort lectures about one of his favourite subjects, 1 June 1976.

just such a painstaking effort. It aimed at approaching the answer to the foremost cosmological question (assuming that the Universe conforms to the Cosmological Principle), viz. - Is the Universe open or closed? - by determining the universal density of visible matter. Oort's careful sums gave an answer, long quoted as a standard result, which fell short by a large factor of the value needed to close the Universe. In 1970 he published a paper entitled *The Density in the Universe* where, further out on a speculative limb than ever before, he theorizes about galaxy formation to conclude that only $\sim$ 1/30 of all matter is in galaxies. If that were so, the density is, within observational errors, equal to the 'critical density' that separates the closed and the open case. Oort seems to very strongly believe that the Universe is balanced precisely on this razor's edge. Whether this belief has esthetic or other grounds I do not know, but it clearly affects the weight Oort attributes to various cosmological investigations. The following footnote to a 1978 Royal Academy contribution may be regarded as his comment on a now-famous comprehensive cosmology paper, two versions of which appeared, one in the Astrophysical Journal, one in Scientific American, some years before:

> *"The density required to make the expansion velocity precisely zero at infinity is called 'critical' density. Some observations appear to indicate that the real density is somewhat lower, but their evidence is contestable; the possibility that the Universe has precisely the critical density remains completely open."*

Is this not a delightful example of how convictions determine our choice of words? In this respect Jan Oort is no exception to the rest of us!

Convictions, contestable though they be, can move us to sound work. If in cosmology the direct determination of universal density is difficult, in practice impossible, there remain many questions amenable to observational scrutiny. The successive appearance of cosmologically relevant structure is a recurrent theme in Oort's more reflective writings, from his 1935 inaugural address onwards. Galaxies - clusters - superclusters, when do they form, in which order and what *primordial* structure do they reveal? How does the population evolution of strong radio sources relate to the galaxy-formation process? These topics are reviewed in a grand sweep in the 1970 Groningen lecture entitled *Galaxies and the Universe*. Earlier I said that this paper is more a programme than a statement closing a career. Oort's bibliography of the 1970's shows that many aspects of that programme have been worked on and a lot of progress has been made. Comparison with an article called Great Expectations published in 1979 demonstrates how specific and heartening this progress has been. We are coming tantalizingly close to the answer of some of the major questions. There is evidence that between redshifts of 0.5 and 1.0 galaxies and galaxy clusters went through major evolutionary phases to attain their present state. That regime comes within reach after Space Telescope is launched in three years! Slightly beyond that, proto-galaxies may be observable in the infrared after IRAS finds them in a deep systematic survey. Evidence for superclustering may be in hand already, as Oort shows in an April 1980 preprint where he gives a daring interpretation of Lyman-α

absorption lines in quasar spectra. How wonderfully rich this decade has been, how tangible the fruits are now of hopes and hints described just ten years ago.

Jan Hendrik Oort, professor of astronomy and director of Leiden Observatory, retired from office as of 1 September 1970. He then continued as an unceasingly active member of the Leiden Observatory research team. I cannot imagine what these ten years would have been without him. His bibliography is only one measure of his contribution. Another is the depth of gratitude and fondness that has grown within us, for encouragement, for counsel, for impatiently urging new initiatives and above all for enthusiasm. Enthusiasm for the phenomena appearing in our telescopes' field of view; enthusiasm for ideas, a contagious convinction that astronomy is important, that the hard work it takes is eminently worthwhile because new and exciting insights, surprising too, are approaching and only just beyond the present horizon. I close this essay with the fervent wish to share Jan Oort's second decade of retirement, a decade in which we may hope to bridge, by observations, much of the gap between $z = 0.5$ and the era of microwave decoupling. Jan, you have kept alive the awareness that all of us together are building towards one great work. Therefore we share your great expectations!

Harry van der Laan did his Ph. D. studies at Cambridge, U.K., was Associate Professor at the University of London, Ontario, and is since 1970 Professor of Radio Astronomy at Leiden.

JAN HENDRIK OORT AND DUTCH ASTRONOMY

H.G. van Bueren

To have studied astronomy under Jan Oort's direct guidance forms an unforgettable experience; to have been a discussion partner in some of his research remains an exalting, be it still somewhat enervating recollection; however, to write on the occasion of his 80th birthday is just an undivided pleasure.

1. Oort's scientific interests and method.

Oort's impact on astronomy is fully based on his scientific work, which has led to a series of important discoveries, and has significantly widened the astronomer's outlook on the Universe. Dutch astronomy has profited greatly from this, and in the wake of Oort's activities the number of astronomers trained in the Netherlands has considerably increased. Oort's own and their successes have placed Dutch astronomy on its present firm basis.

When Oort began his work in the early twenties, astronomy in Holland was dominated by Kapteyn, who had just completed his pioneering work on the structure of the Milky Way, and who guided Oort's first steps in Groningen. He and De Sitter at Leiden have both strongly influenced the course of Oort's future work.

The prospects for astronomy in the Netherlands were much less bleak in 1920 than they were when Kapteyn started his career in 1875. Contacts with the international astronomical world had in the meantime been firmly established by Kapteyn, and would henceforth strongly influence the work at all Dutch observatories. The most burning problem of the time, that of the scale of the Universe, thus naturally became a focal point of interest for Oort. His later contributions to its solution - still not fully obtained -, for instance by his demonstration of the differential rotation of the Galaxy, have been fundamental.

One of the most significant aspects of Kapteyn's work was his demonstration that by application of statistical methods significant conclusions could be drawn as to the structure and state of motion of a system of a large number of stars, such as the Galaxy. Conscientious determination of the mean and the deviation from the mean led to surprising possibilities! Kapteyn's model of the Milky Way, however, was

H. van Woerden, W. N. Brouw, and H. C. van de Hulst (eds.), Oort and the Universe, 31–37.

later proved wrong, and even in his own time it was already clear where it might go wrong: there was no absolute clarity about the role of interstellar absorption, and there were inconsistencies in the dynamics of the model. The latter were first noted by Oort, for instance when he realized that the "Kapteyn universe" did not contain enough mass to bind the high-velocity stars to it.

In retrospect it is surprising that Oort nevertheless did not lose trust in the statistical methods developed by Kapteyn. On the contrary, he succeeded in applying them even more judiciously. This certainly shows great scientific insight or intuition, and presumably both.

Oort has remained interested in the statistical significance of stellar proper motions up till now. The results of his patient and inflexible attention to the forthcoming availability of new and reliable data usable for renewed determinations of the parallax of the Hyades, of the constants of differential galactic rotation, A and B, of the famous K_z, or of the distance to the Galactic Centre, to name but a few examples, are well known and have served in their time as excellent sporting grounds for younger colleagues as well as old friends. This early focus of Oort's interest has set the primary line of research in Leiden for many, many years.

His researches on the broad subject of the structure of the Universe were thus started as it were from the inside, by studying the distribution and motions of stars and gas clouds in the neighbourhood of the Sun. He went on by unravelling the spiral structure of our Galaxy, using radio-astronomy. Still later, he has concentrated on external galaxies, although he remains continually occupied with the Galactic Centre also, studying this from such diverse observational angles as RR Lyrae variables, planetary nebulae and (explosive) cloud motions. His well-known attempt (1970) to determine the density of the Universe may well have been originally inspired by De Sitter, and forms an example of his interest in cosmology, which was later strongly stimulated by the fine observations obtained with the Westerbork Synthesis Radio Telescope in the seventies. With ongoing years Oort's field of interest has forever widened, ranging from the enigmatic Galactic Nucleus through the presence and properties of M-dwarfs in the Galaxy and the problem of the hidden mass to large-scale galaxy counts and the theoretical question of the origin of galaxies. It is probably not so surprising that it returns also now and again to old subjects such as the Hyades and the velocity distribution of faint stars!

In many of the modern cosmological researches the presence of an element of speculation cannot be denied. Oort's speculations are not entirely absent, but they can always be traced back in a very few steps to reliable observations, and are never fantastic. Dutch astronomy is not fantastic either, but rather characterized by its sobriety. I believe that this is one of Oort's good influences. He draws a very strict line between speculation and fantasy, stricter than many of his colleagues abroad - whom he holds in high esteem, nevertheless, just because of their daring!

Most instructive for all students who have followed Oort was the way he proceeds in obtaining his results from raw data. The extreme care and precision, and the absence of any noticeable hesitation in rejecting the work of months and starting all over again when new and better data become available, sets to them an impressive and enduring example of the conscientious scientist. Moreover, in progressing he notes successively every stage where the situation still needs some clarification, and tackles, or persuades others to tackle, these bypassed problems in due course - which may mean many years later. In this way Oort succeeds again and again in slowly but surely building up a coherent mass of fact and theory, which finally yields a result as reliable as can be within the limits of astronomical observational accuracy.

This manner of work is reflected in his writing - Oort's papers contain much text and usually rather few formulae; the wording is carefully chosen, and the reader is gently but compellingly drawn in the direction Oort wants him to go. Oort's papers are not "difficult" in the way many modern authors write, but they are extraordinarily full of content, which makes them anything but easy to read.

2. Oort's influences on Dutch astronomy and astronomers.

Scientific method.

The effect of Oort's characteristic scientific method on his Dutch colleagues and students is one that should be there to last. This effect, of course, is not general, but for those who have felt it, it is of great value. It is also in a way unsettling - not many of us can withstand the seductive possibilities of jumping to conclusions to save time, something that Oort abhors.

Oort's papers are for a large part written solely by himself. Obviously he refrains from adding his name as a co-author to the work of his disciples, although his influence on that work is always unmistakably there. Now and again he asks somebody to work with him on certain details, and then two or sometimes even three names appear. But his method is essentially an individual method, and the number of his pupils is not remarkably high. He usually does not take more than one at a time, and he takes his time. Many of his pupils have also become individual, not extremely prolific authors, another sign of the powerful influence of his method.

Choice of problems.

However, Oort exerts his influence not only by his way of doing things, but also by what he does. We have already mentioned his contributions to the development of the field of stellar dynamics, where Oort, together with B. Lindblad, solved one of the problems unsolved by Kapteyn. The other problem, that of interstellar extinction, did not escape his attention either. Here a complication arose. A detailed investigation into the physical processes responsible for absorption and scattering by atoms and particles Oort felt to lie outside his region of competence. In such a case he never hesitates to consult others, and in the present instance the theoretical physicists Kramers and Ter Haar, and somewhat later the theoretical astrophysicist Van de Hulst were drawn to his aid. Together with them (and undoubtedly leading them on) he contributed to the foundations of the theory of interstellar extinction, underlining especially the significance of dust grains. This subject would later become another of the main lines of astronomical research at Leiden, where it ultimately led to the Department of Laboratory Astrophysics, as well as at Groningen, where it finally resulted in the infrared-astronomy branch of the Kapteyn Institute, of which at present the IRAS satellite promises to become the most spectacular product.

One cannot conclude from this, of course, that Jan Oort has founded laboratory astrophysics and infrared astronomy in Holland. No, but he has shown at an early date that problems existed in astronomy that could not be fully solved with the conventional means available at the time. In a particularly Oortian way such problems were dished up to colleagues, who were "used" to bring them forward. In using others and especially colleagues, Oort is great. We are all thankful to him for this, since invariably the "used" have profited at least as much as the "user"!

Once it was established that there is interstellar absorption and also that it can be explained, Oort characteristically changed his interest from this slightly "too physical" subject back to astronomy proper. The interstellar medium remains of interest to him, of course, but from now on it becomes the gas that fascinates him in particular. Everyone knows now how the neutral hydrogen atoms making up most of this gas show their presence through the hyperfine radio transition at 21 cm wavelength, but it has largely been by Oort's continuous pressure and his persuasion of others (most notably Van de Hulst) to study particularly aptly chosen details, that hydrogen-line radio astronomy was actually born. This and the subsequent expansion of general radio astronomy in the Netherlands may well prove to be the most powerful concrete product of Oort's many influences. Since radio astronomy will be fully discussed elsewhere in this book, we shall refrain from elaborating on it, but use it here only as an illustration of the sometimes uncanny _intuitive power_ of Oort. His ability to define problems and subjects of research that are not only solvable but above all significant, and that will lead to further important research - a well-known and most important criterium for good research policy as first formulated by Weinberg-, is truly surprising. For example, a problem that interests him strongly for a number of reasons and to which he has directed the attention of a number of his pupils and colleagues, is that of the Crab Nebula. It hardly needs mentioning here how many and diverse the consequences of the observational and theoretical studies of this supernova remnant have proved to be. The origin of most of these studies can be easily traced back to Oort.

In these ways most, if not all, of the present Dutch astronomers have one or more times profited in some way or another from Oort's suggestions and ideas, even when no larger-scale consequences can (yet) be indicated. He is wont to present many of his new thoughts already in an early stage to the astronomical community at the meetings of the Nederlandse Astronomen Club, and to his colleagues learned in other fields in the sessions of the Royal Netherlands Academy of Sciences. His lectures, soft-toned and sometimes hardly audible, but always clearly and systematically composed, are invariably full of new ideas and lead the listeners right up to the boundaries of astronomical knowledge. More often than not they even give them a glimpse of what might lie beyond.

As an observatory director and professor.

A part of astronomy to which Oort does not care to contribute directly is the astrophysics of stellar structure. Nor is he actively interested in solar physics. Another boundary is apparently drawn where space research proper is concerned. These fields he leaves for active exploration to others, colleagues at Leiden and Groningen as well as Utrecht and Amsterdam. He admires the results obtained in these fields. but seems to hesitate to become involved. This has led to a certain separation and division of interests between the Dutch observatories, which may certainly be called an influence of Oort. During his directorate of Leiden Observatory, "astronomy" has coincided with "Leiden" in Oort's mind. Far from deprecating the work done elsewhere in the country, however, he followed that attentively, and in most cases where it proved successful or yielded promising young researchers, he was keen to profit for Leiden in one way or another. In this respect also, Oort's presence was felt throughout the country and in all its observatories. In many cases, and particularly through his influence in the I.A.U., he has stimulated and enabled budding Dutch astronomers to form international contacts, and in this way contributed much to their scientific education.

Oort has used his directorship and professorate at Leiden to make this institute a truly international visiting centre for many foreign colleagues as well as for a large number of his Dutch friends and for students of other universities. In close co-operation and fairly strict division of work and interest with his senior colleagues and deputy directors, he maintained there a high standard of intellectual proficiency (while simultaneously promoting other social aspects like rowing, modern art and ice skating), and succeeded in inspiring, in his quiet but irresistible way, all his students and co-workers to perform to their utmost capacity. He knew well how to choose the right person for the right problem, whether scientific or organisational. It was not always easy to work in the daily presence of such an urbane but strict and exacting master, who moreover towered above his pupil in insight and - complete - knowledge of literature.

Oort's method of instructing and directing is not based on a specific technique, nor is it philosophically founded, but it can best be described as "based on observation". He therefore has hardly ever fallen into the trap of his own theory - new observations generate new ideas in his mind and he does not hesitate to replace older ideas with them. This capability underlies the strong grasp he has had on the work of all his pupils and co-workers. But it is not the fastest way that leads to the formation of a "school", and indeed one cannot speak of an existing school of Oort's pupils. Maybe one should rather consider the whole of present Dutch astronomy as some sort of "school", since this astronomy is strongly based on continuous confrontation with the latest in observations - clearly no little due to Oort's working.

Another influence of Oort, not very tangible but surely present, is exerted by his willingness to read and judge the work of others, a willingness which mostly results in an extension and deepening of that work, with the help of the numerous recent preprints that Oort always seems just to have received.

A discussion with Jan Oort makes his personal influence fully felt. The hours of "Oortian silence" interspersed with carefully phrased questions and statements, will remain an unforgettable experience for everyone, as will his minutely written notes on small pieces of paper.

As a national scientist and negotiator.

Outside his Observatory, the first impression one forms of Oort is his modesty. He seems not to want to influence his colleagues, and it appears as if, when he does, he does so reluctantly. Nevertheless, just by this influence he has shaped a large part of Dutch astronomy, in and outside Leiden. He does such shaping in official and non-official discussions, where one of his most remarkable traits, namely his tenacity, sometimes even obstinacy in matters he thinks right, plays an important role. The kind and courteous Oort holds fast to his opinions with unbelievable firmness. He therefore finally always wins his argument, with colleagues as well as with research-financing organisations and government officials, and most Dutch astronomers, although at first sometimes reluctant, are after all grateful to him for this, because in most cases he has indeed proved to be right.

In a small country like the Netherlands, the presence of a dominating top scientist such as Jan Hendrik Oort naturally has consequences for all those active in his science. Concepts and interpretations now appearing self-evident to them were first phrased and proved by Oort, learnt under Oort's guidance or taught in his courses. Also organizationally much has been changed by him. Comparing the situation in Dutch astronomy in 1980 with that of 60 years ago, one sees this quite clearly. The present field is strewn with monuments erected or initiated by Oort: the Netherlands Foundation for Radio Astronomy and its Synthesis Radio Telescope, the journal Astronomy and Astrophysics, the European Southern Observatory, to name but a few. Most of these monuments are international, because Oort, in the true spirit of Kapteyn, has first and foremost remained an international astronomer. He has led the Netherlands up in the astronomical world during 60 years by opening up a number of important new pathways, and above all he has taught them the fundamental importance of both conscientiousness and patient carefulness in all sorts of research.

Since his retirement he is as active as ever, as will be described elsewhere. Astronomy has gone her own way since then and has more and more become big science. Oort has taken part in the beginning of this development, but he has never allowed organization to dominate over science. One wonders whether this will not prove the most important one of the many lessons he has taught us.

H.G. van Bueren studied physics and astronomy at Utrecht and Leiden, was assistant at Leiden Observatory from 1946 to 1950 and obtained a doctorate in solid-state physics. He was Professor of Solid-State Physics from 1962 and Professor of Astrophysics at Utrecht from 1964 till 1980, and is now chairman of the Science Policy Advisory Council in The Hague.

OORT'S SCIENTIFIC IMPORTANCE ON A WORLD-WIDE SCALE

Bengt Strömgren

On contemplating the title for a contribution to the "Liber amicorum for Jan H. Oort" suggested to me by the Editors, I tried to imagine the situation of a historian of natural science, working some time during the twenty-first century on a monograph describing Jan Oort's scientific importance. If successful, this historian would have covered in his monograph a very substantial portion of the history of astronomy in our century.

I must limit myself to select examples illustrating what Jan Oort's scientific work, his ideas and his inspiration, have meant to astronomers all over the world, and to the development of astronomy.

Jan Oort's thesis (1926) was entitled "The Stars of High Velocity". It is wellknown how his novel approach led him to findings published in papers culminating with "Dynamics of the Galactic System in the vicinity of the Sun" (Oort, 1928). In an article in this volume Per Olof Lindblad has described in detail how the pioneering investigations of Jan Oort and Bertil Lindblad progressed. Through their efforts and results a whole new framework for galactic research, in particular on galactic dynamics, was created.

In the 1920's the role of light-absorption by interstellar matter was not generally appreciated. However, it can be inferred from Jan Oort's writing that he fully realized its importance even before the availability of the observational evidence convinced the astronomical community during the 1930's.

During the 1940's Jan Oort turned his attention to problems of interstellar matter. In his George Darwin Lecture (Oort, 1946), entitled "Some phenomena connected with interstellar matter", he gave an overview of the subject and presented a number of new results obtained by himself and his collaborators. When one reads this paper today, one is struck by the fact that a number of problems clearly formulated and discussed by Jan Oort in this lecture became the subject of intensive research during the following decades. Here are discussions of the picture of interstellar clouds and intercloud matter, of the relation between gas and dust, of the effect of collisions between clouds, and of the build-up and destruction of interstellar particles. The interaction between expanding supernova shells and surrounding interstellar matter

H. van Woerden, W. N. Brouw, and H. C. van de Hulst (eds.), Oort and the Universe, 39–44.

(as illustrated by the Veil Nebula in Cygnus) is discussed in some detail.

Jan Oort had collaborated with J.M. Burgers on the question of collisions between interstellar clouds, and this collaboration led to the organisation of a Symposium in Paris (1949) on the motion of gaseous masses of cosmical dimensions, jointly sponsored by the International Union of Applied Mechanics and the IAU. This symposium was followed by others on the same general subject, symposia which together had a considerable influence on the development of cosmical aerodynamics.

In 1954 and 1955 Jan Oort returned to the subject of the dynamics of interstellar clouds with the papers "Outline of a theory on the origin and acceleration of interstellar clouds and O associations" (Oort, 1954) and "Acceleration of interstellar clouds by O-type stars" (Oort and Spitzer, 1955). During the following decades the area opened by these researches developed into the broad field of research concerning star formation and interaction between stars and interstellar matter.

The study of the properties of interstellar matter had become a significant subject, closely related to others of importance in astronomy and astrophysics. The fact remained, however, that the very existence of this strongly light-absorbing medium presented formidable obstacles to exploration of the Galaxy.

In an article in the volume "Onder de ZWO-Bannier" (dedicated to J.H. Bannier on the occasion of his seventieth birthday) Jan Oort (1979) has described how the pioneer work of Grote Reber - that led in the 1940's to the first charting of radio-radiation originating in the Galaxy - made it clear that radio astronomy held great promise for galactic research. The radio waves in question would not be absorbed by interstellar matter and could penetrate the whole length of the Galaxy. Furthermore, the perspectives widened greatly in view of the prediction in 1944 by H.C. van de Hulst (1945) that the 21-cm line emitted by interstellar neutral hydrogen should be observable with the help of radio telescopes.

Plans developed in the Netherlands for the construction of a 25-meter radio telescope that would be chiefly used for the exploration of the Galaxy through observations of the 21-cm hydrogen line.

From certain points of view the project must have appeared a difficult one. In the Netherlands there were not - as in Great Britain, Australia and the United States - groups of scientists who had worked during the war years on the development of radar and other communication technology. In the article just mentioned, Jan Oort mentions a number of fortunate circumstances which helped to make the project successful in the end. There is little doubt, however, that the main reason for success was Jan Oort's own contribution.

The 25-meter radio telescope at Dwingeloo went into operation in 1956. In the meantime, however, radio astronomy in the Netherlands got a start with the 7.5-meter radio telescope in Kootwijk.

As is well-known, the year 1951 brought a discovery of the greatest importance to the development of the new field of radio astronomy. The prediction by van de Hulst was confirmed when the first successful observations of the 21-cm line emitted by galactic interstellar gas were made, first by Ewen and Purcell (1951) at Harvard, then by Muller and Oort (1951) in the Netherlands with the 7.5-meter radio telescope, and by Christiansen and Hindman in Australia. In December 1951, Jan Oort delivered the Henry Norris Russell lecture. His title was "Problems of galactic structure". He stated that he had on this occasion set a task for himself, namely, "to take stock of what knowledge has been acquired from the visible wavelengths and to compare this with the tentative first observations of the new era", an era characterized as the ascendance of radio astronomy.

The lecture (Oort, 1952) contains a wealth of new results: A detailed model of the Galactic System, determinations of the distance from the Galactic Centre to the Sun and of the circular velocity of rotation at the distance of the Sun. Further, an estimate of the optical depth per kiloparsec in the 21-cm line and of the average density of interstellar neutral hydrogen in a region of about 1 kiloparsec radius around the Sun. The possibility of deriving the circular velocity of rotation in the Galactic System as a function of distance from the centre is discussed in detail and first results pertaining to this problem, obtained by C.A. Muller with the 7.5-meter radio telescope in Kootwijk, are presented.

There is no doubt that this contribution by Jan Oort had a profound influence on the attitude of astronomers all over the world toward the new discipline of radio astronomy, and that it greatly helped its development during the years thereafter.

In the Netherlands the 7.5-meter radio telescope was used to chart the sky in the 21-cm line. There followed the demonstration by Jan Oort and his collaborators that interstellar neutral hydrogen is concentrated in relatively narrow arms, analogous to the spiral arms observed in extragalactic systems. In June 1954 Jan Oort, Henk van de Hulst and Lex Muller published their paper "The spiral structure of the outer part of the galactic system derived from the hydrogen emission at 21-cm wavelength" (Van de Hulst, Muller and Oort, 1954).

After the 25-meter radio telescope in Dwingeloo and its equipment for making 21-cm line observations had become operational, Jan Oort turned his attention to the central regions of the Galaxy. In collaboration particularly with G.W. Rougoor the expanding 3-kiloparsec arm was discovered and investigated, as was rapidly rotating neutral hydrogen gas closer to the Galactic Centre. A general tendency of expansion of gaseous matter away from the centre was found.

Here we have another example of Jan Oort's investigations opening up a new field that became very active and to which astronomers in a number of countries made important contributions. Jan Oort continued his work on the problems of the central region of the Galaxy, combining results from radio and optical as well as infrared astronomy, his efforts culminating, so far, with his article, "The galactic center", in the Annual Reviews (Oort, 1977).

The discovery and investigation of high-velocity clouds falling into the Galaxy was another notable contribution made with the 25-meter Dwingeloo radio telescope. The subject has continued to develop and presents challenges in both galactic and extra-galactic research.

A radio-continuum survey at relatively high frequency made with the Dwingeloo telescope by G.Westerhout led to the identification of HII regions, many of which were invisible to the optical observer because of interstellar absorption. This work had considerable influence on the following development of radio-astronomical galactic research, in particular concerning problems of star formation.

During the 1950's there occurred a striking development in radio astronomy in areas other than 21-cm hydrogen line research. By 1954 about one hundred cosmic radio sources were known. The positional accuracy was about 10 minutes of arc, and this accuracy was sufficient for the first few certain identifications with optical counterparts. There followed a breakthrough with regard to positional accuracy, which improved by more than an order of magnitude, with consequent successes of identification with optically known galactic objects and - particularly - external galaxies. Radio astronomers in Great Britain, the United States and Australia had played a leading role in this development.

The 25-meter Dwingeloo radio telescope had been constructed particularly for the purpose of galactic research. In 1959 Jan Oort and his collaborators in the Netherlands began to develop plans for a large radio telescope, particularly suited for extragalactic research. Charting of limited areas with an angular resolution of 20 seconds of arc was aimed at, and the radiation sensitivity was to match the resolving power, i.e. be sufficient for recording extragalactic sources expected to come within reach when charting was done with the resolution mentioned. The result of strong and sustained efforts was the Westerbork Synthesis Radio Telescope - WSRT - dedicated by Queen Juliana in 1970. The WSRT is referred to in many places in this volume, and the story of its development is here told by Jan Oort's close collaborator on the project, W.N. Christiansen.

Westerbork has become a very important centre. Scientists from half-a-dozen countries that are particularly active in radio astronomy, work regularly there and take advantage in their research of the outstanding qualities of the installation. And the successes in radio-astronomical research of the Westerbork installation have influenced the development of radio astronomy in other countries, particularly the United States.

Consider the development of the techniques of identifying optical counterparts of radio sources, so important because it makes possible the combination of observational data from a very wide range of wavelengths. In 1954, the positional uncertainties amounted to 10 minutes of arc, with corresponding large "error boxes" on the identification photographs. Today, Westerbork charts yielding positions accurate to a few seconds of arc have been compared with deep plates obtained with the new generation of 4-meter reflectors. The resulting identifications are based on coincidence in position alone, and thus yield unbiased samples.

Consider the charting at Westerbork of the nearer spiral galaxies such a Messier 51 and Messier 81 in the synchrotron-continuum and the 21-cm line, respectively. The interpretation of these observational results in terms of the wave theory of spiral arms meant a very considerable extension of the scope of galactic dynamics.

Consider the cases where the charting of galaxies by the WSRT has revealed effects of gigantic explosions originating in galactic nuclei. Here we see contributions to the highly active field of extragalactic research which is concerned with explosive galaxy nuclei.

Consider finally the mapping by the WSRT of galaxies with tails that have been modified because the galaxy is moving through intergalactic matter, and we note an important contribution to the developing field of study of interaction of galaxies and intergalactic matter.

In his Henry Norris Russell lecture on problems of galactic structure, Oort (1952) spoke of the comparison of results acquired from observations at visible and at radio wavelengths. Let us now return to optical astronomy - to Jan Oort's contributions in this field.

During Walter Baade's stay at Leiden Observatory as a guest professor in 1953, a plan was born to create a European Observatory in the Southern Hemisphere. The background is well known. The European countries had, in observational astronomy, been lagging farther and farther behind the United States.

In 1953 Jan Oort invited astronomers from Belgium, France, Great Britain, Sweden and West Germany to a meeting at Leiden for a discussion of the plan for a European Southern Observatory. This was the beginning that led to the creation of the European Southern Observatory organization - ESO.

Jan Oort had invited me to participate in the meeting, although I had at that time left Denmark and worked in the United States. I travelled from the meeting in Leiden to Copenhagen with Bertil Lindblad. We were both deeply impressed with the plan for ESO, and having arrived together in Copenhagen we decided to visit Niels Bohr in his summer home in Tisvilde in order to discuss the proposed plan with him. Niels Bohr's very positive reaction to the plan strengthened our belief in its excellence and feasibility.

During the following years Bertil Lindblad, André Danjon and Otto Heckmann worked closely with Jan Oort toward the realization of the planned European Southern Observatory. It took a long time until a treaty could be signed, but in 1962 the European Southern Observatory came into existence, supported by five member countries: Belgium, France, the Netherlands, Sweden and West Germany. Denmark became ESO's sixth member in 1967.

The story of the development of ESO into a flourishing organization has been told in other contexts. Suffice it here to mention that Jan Oort's contribution to its creation and development was all-important. His clear view of the scientific goals, and his insistence that ESO contribute not only through its observational facilities but also by creating a more favourable climate for European collaboration on astronomical problems, meant very much indeed.

In the volume "Onder de ZWO-Bannier", to which I referred before, Jan Oort has paid tribute to J.H. Bannier and emphasized how much his work for ESO meant. Surely, Jan Oort had outstanding collaborators in ESO matters. I am convinced, however, that Jan Oort meant more than anyone else, and that here is still another splendid example of Jan Oort's scientific importance on a world-wide scale.

Let us now return to the problems that historians of natural science encounter, and imagine not a person working in the twenty-first century, but someone working at a much later time. Suppose that he had first studied the development of mathematics in the twentieth century, and that during these studies he would have discovered that the mathematician Bourbaki was not a real person, but a name covering a whole group of mathematicians. Working on the development of astronomy in the twentieth century and faced with the extent and importance of Jan Oort's contributions, this historian might well be tempted to advance a theory inspired by the Bourbaki story.

We, Jan Oort's friends, know better. To us, Jan Oort has been, and is, always the same wonderfully inspiring, kind and true friend. On the occasion of his 80th birthday I, and his many, many friends, wish to thank him most sincerely, and to tell him how much we are looking forward to coming years of continued leadership by Jan Oort.

References

Ewen, H.I. and Purcell, E.M. 1951, Nature 168, pp. 356.
Muller, C.A. and Oort, J.H. 1951, Nature 168, pp. 357.
Oort, J.H. 1926 , "The stars of high velocity", thesis, Groningen University.
Oort, J.H. 1928, Bull. Astr. Inst. Netherlands 4, pp. 269.
Oort, J.H. 1946, Mon. Not. Roy. Astron. Soc. 106, pp. 159.
Oort, J.H. 1952, Astrophys. J. 116, pp. 233.
Oort, J.H. 1954, Bull. Astr. Inst. Netherlands 12, pp. 177.
Oort, J.H. 1977, Ann. Rev. Astron. Astrophys. 15, pp. 295.
Oort, J.H. 1979, in "Onder de ZWO-Bannier", ZWO, Den Haag, pp. 85.
Oort, J.H. and Spitzer, L. 1955, Astrophys. J. 121, pp. 6.
Van de Hulst, H.C. 1945, Ned. Tijdschr. Natuurk. 11, pp. 210.
Van de Hulst, H.C., Muller, C.A. and Oort, J.H. 1954, Bull. Astr. Inst. Netherlands 12, pp. 117.

Bengt Strömgren studied and worked at Köbenhavn Observatoriet, then at the Yerkes and MacDonald Observatories. He was General Secretary of I.A.U. in 1948-52, and I.A.U. President in 1970-73. He was President of the Council of the European Southern Observatory from 1975 to 1978 and is now Director of the Nordisk Institute of Theoretical Astrophysics in Köbenhavn.

Bengt Strömgren

OORT AND INTERNATIONAL CO-OPERATION IN ASTRONOMY

D.H. Sadler

My first sight of Jan Oort was at the evening reception at l'Observatoire de Paris in July 1935 on the occasion of the fifth (and my first) General Assembly of the International Astronomical Union (IAU). At that meeting he took over the duties and responsibilities of the office of General Secretary from F.J.M. Stratton (who, incidentally, had first interested me in astronomy in Cambridge).

The physical comparison between Stratton's rotundity and Oort's slightness of figure was remarkable, and was duly remarked upon. The young British astronomers, who were fully aware of Stratton's outstanding qualities, could not then have appreciated that their external differences concealed equally dedicated, efficient, accomplished and charming personalities.

I next saw Oort, but do not recall actually talking with him, at the sixth General Assembly of the IAU in Stockholm in August 1938. Astronomy, which in most countries was inadequately financed, was advancing quietly but the IAU continued to progress in both size and stature. Since 1935 it had grown in membership by 25 percent (a modest 400 to 500), a rate of "inflation" that (with fluctuations and apart from the war years) continued for many years. There was a significant increase in the numbers and extent of co-operative programmes and, in spite of some continuing difficulties in respect of national membership, almost all astronomers were involved. The General Assembly was a happy and successful meeting, but the prophetic words of the newly-elected President, A.S. Eddington, in his speech of acceptance were only too soon to be fulfilled.

> "We cannot foresee what may happen before we meet again in "Zürich. On the astronomical side we can make some guesses. The 200-inch reflector will be completed or approaching completion.... But on the international side no one dares prophesy. But, if in international politics the sky seems heavy with clouds, such a meeting as this at Stockholm is as when the Sun comes forth from behind the clouds. Here we have formed and renewed bonds of friendship which will resist the forces of disruption".

H. van Woerden, W. N. Brouw, and H. C. van de Hulst (eds.), Oort and the Universe, 45–50.

Alas! The war brought the formal activities of the IAU to an abrupt end, although astronomers throughout the world continued to communicate and co-operate informally. The President died at a tragically early age in 1944 and was succeeded by H. Spencer Jones; in the meantime the General Secretary endured the restrictions and hardships of those in countries occupied by enemy forces. My first meeting with Jan Oort that I specifically recall was at the Royal Astronomical Society (of which I was then Secretary) during his first visit to England, after the war, in June or July 1945. Then, and later, he said little about the war years; but it was quite clear to us that the difficulties, with which he had to contend, had been enormous.

However, national and international conditions improved sufficiently to allow the seventh General Assembly of the IAU to be held in Zürich in 1948. I was appointed to serve as the U.K. representative on the Finance Committee, and was duly elected on the working sub-Committee as vice-chairman; the chairman, C.S. Beals, had to return to Canada before the final session. For the first time I became aware of the whole range of the duties and responsibilities of the General Secretary (the IAU has no separate office of Treasurer), and I vividly recall my unstinted admiration for the manner in which the complex administration of the Union had been conducted. But far more important was the fact that the astronomical activities of the Union, including many co-operative projects, were again in operation and that relations between astronomers in all countries had been renewed.

The aftermath of war had inevitably left some astronomers with reservations about the desirability of a full resumption of friendship with others. However, German astronomers were invited to Zürich on a personal basis and both they, and many astronomers from the U.S.S.R., made valuable contributions to the meetings and to international programmes. Much of the credit for this encouraging state of affairs and for the extremely successful Zürich meeting, both in scientific and personal matters, must go to the energetic endeavours of Oort, in full co-operation with Spencer Jones, in the few years after the end of the war. At the final session of the General Assembly he received a well-deserved tribute from the President, and a standing ovation from all. A future President (Otto Struve) wrote:

> "His quiet, efficient, and impartial conduct of the work of the Union during the difficult years of the war, and especially during the past three years of reconstruction, has not only won him the admiration, respect, and affection of the vast majority of astronomers of all nations, but has created a lasting monument in the history of international science".

Oort had served 13 years as General Secretary.

My next substantial meeting with him was in Leningrad in May 1954 when we participated, as representatives of our respective countries, in the celebrations marking the reconstruction of Pulkova Observatory, which had been completely destroyed during the war. That occasion did much, in my opinion, to pave the way for the invitation, and its acceptance, to hold the tenth General Assembly of the IAU in Moscow in 1958; and Oosterhoff and Oort were largely responsible for ensuring that the cancellation of the proposed General Assembly in 1951 (which did not, of course, affect the personal relationships between astronomers) was forgotten. One personal note, giving an insight into Oort's friendliness and consideration (as well as his determination and persistence) might be mentioned here. The interpreter, Tanya, assigned to him by the Academy of Sciences, was a most efficient and helpful person who went far beyond her duties in assisting him with the preparation of the addresses and papers that he presented during the various meetings; in 1958, after some effort, he traced her home address and went to see her.

As General Secretary of the IAU Oort had attended meetings of the Executive Board of the International Council of Scientific Unions (ICSU), which was then beginning, in line with the activities of UNESCO, to expand its interests and activities beyond those of the individual constituent Unions. He maintained his interest after 1948, although he was not again a member of the Board until he became President of the IAU in 1958 - when he (together with myself as General Secretary) attended the meetings of the Executive Board and the General Assembly of ICSU in Washington in October. With his broad scientific knowledge and judgement, his previous experience and (let it be said) his high international standing, he was able to exert considerable influence on the policies and activities at a critical time of expansion and change. The International Geophysical Year (IGY) project was being extended, and the Committee on Space Research (COSPAR) initiated. Both then, and at subsequent meetings of the Board in 1959 and 1960, Oort made many significant contributions to the work of the ICSU. However, perhaps the one that may have given him the greatest satisfaction, and astronomy its most lasting benefit, is the establishment of IUCAF (Inter-Union Committee on the Allocation of Frequencies - for Radio Astronomy and Space Science), through which the requirements for scientific research can be authoritatively presented, in competition with the many other interests involved, at the periodic Administrative Conferences on Frequency Allocations convened by the International Telecommunication Union (ITU). Earlier it had been due to his initiative, energy and personal persuasion that the IAU, in co-operation with URSI and COSPAR, was able to secure, at the ITU Administrative Radio Conference in 1959, both recognition of the need to reserve frequencies for scientific research and significant allocations.

I well recall his telephone call from Leiden in which he emphasised, in his quiet way, the urgency and importance of ensuring that a representative be present in Geneva throughout the whole three months of the conference. The IAU's budget made no provision for such expenditure but we decided, there and then, to go ahead and to seek subsequent approval for our action from the Executive Committee and the General Assembly. In fact the actual cost was relatively small, largely because of the generous contributions made by the observatories and establishments that seconded members of their staffs to watch over the astronomical interests. Oort himself spent some time in Geneva talking to delegates; he presented an evening discourse on Radio Astronomy to the (national) representatives who were mainly concerned with the demands for broadcasting, and other radio, services to the public. He must have made a considerable impression on his audience, as many later expressed understanding and support.

In 1957 I was invited by the President of the IAU, A. Danjon, to act as assistant to the then General Secretary, P.Th. Oosterhoff (who had himself acted in a similar capacity in 1951 and 1952). In due course I made my first visit to Leiden where, during our detailed discussions, we had the encouragement and advice of Oort, who maintained a close interest in the organization and administration of the Union, as well as a comprehensive involvement in a vast range of its scientific activities. He was for many years an active member of no fewer than nine Commissions, played a leading part in the General Assemblies in 1952 (Rome) and 1955 (Dublin) and participated in many Symposia and Joint Discussions. About this time Oort was also much engaged with the preliminary organization of the European Southern Observatory (ESO) project, which he did so much to initiate and press through to its successful outcome. I recall his disappointment that, in spite of initial support for the project from some British astronomers, the U.K. eventually decided not to participate. He had been General Secretary in 1946 when, in the aftermath of a destructive war, the proposal for setting up an international observatory was first suggested (jointly by H. Shapley and the Polish astronomers); this suggestion led to the formation of an IAU Commission on the subject in 1947, to be formally confirmed by the General Assembly in 1948. Although by no means the first observatory to be shared by astronomers from several countries, ESO was the first to be planned specifically for the purpose, and its establishment must have given Oort (and all others concerned) great satisfaction, in giving practical expression to the Objects of the IAU and the furtherance of astronomy.

At the tenth General Astronomy in Moscow in 1958 Oort was elected President of the IAU while I succeeded Oosterhoff as General Secretary. We were immediately faced with a difficult problem arising from the interference of politics with the facility, which all astronomers need and want, to co-operate fully and freely with their colleagues in other countries. As twenty-two years later there now seems to be a possibility of welcoming back Chinese astronomers as Members of the Union, there is little to be gained by recalling the circumstances that led to China's withdrawal from the Union. But Oort's firm insistence on strict adherence to the Statutes, and his complete integrity, while not preventing the inevitable consequences of political discord, did prevent internal dissension within the Union. His wise judgement in such matters was fully used by the ICSU in the many discussions that took place at meetings of the Executive Board, and special committees, on the questions of political discrimination that arose among the scientific communities.

It was a great privilege and pleasure to work closely with Jan Oort: a privilege to benefit from his depth of understanding and experience, and a pleasure to have his views expressed with so much precision and clarity. In addition to the not inconsiderable "business" of conducting the affairs of the Union, the three years 1958-1961 brought a number of special problems affecting the organization of the Union or broader issues of international co-operation. One of the most interesting, and far-reaching, changes concerning the structure of IAU Commissions illustrates Oort's ability to obtain a consensus from a broad range of initially disparate views. At his invitation a small committee of the Executive Committee, together with J.C. Pecker and E. Schatzman who had jointly submitted a series of somewhat radical proposals, met in Leiden in the spring of 1960. It took two full days of discussion, under the patient and understanding chairmanship of Oort, to arrive at a series of recommendations that, I think, satisfied everyone - and still, after twenty years, continue to serve their main purpose.

In general the internal affairs of the IAU are secondary to its overall object of furthering international co-operation. But occasionally a particular action may have a special effect: in 1959 Oort took the view that it would be a helpful gesture to hold a meeting of the Executive Committee in one of the (astronomically) smaller countries, and suggested that B. Sternberk (then a Vice-President) arrange for an invitation to be issued on behalf of Czechoslovakia. That meeting was an outstanding success. The visit undoubtedly gave encouragement to the Czech astronomers, and members of the Executive Committee were much impressed by the astronomical work being done with relatively modest resources. The policy of holding scientific and other meetings, where practicable, in the smaller countries has since been continued.

Reference has been made to the conflict between astronomy and society in respect to the allocation of frequencies for radio astronomy. Another example was the project, known as West Ford, of placing a reflecting screen of small dipoles in orbit round the Earth. Oort dealt with both of these matters in the course of his masterly Presidential address at the eleventh General Assembly in Berkeley in August 1961. The Executive Committee acted with determination, and two strong resolutions were adopted by the General Assembly; these stated unequivocally the IAU's attitude towards space projects and experiments that might lead to contamination of space with adverse effects on astronomical research. An interesting side-light emerged: the main Resolution (notwithstanding protocol, since the IAU is a non-governmental organization) was addressed to all Governments and I, as General Secretary, was requested to communicate it direct to the Heads of State; in spite of my strong reservations regarding this procedure, a surprisingly large number of countries expressed positive support, and many replies were personally signed by the Head of State!

Oort's Presidential address, delivered in English and French, contains many examples of the qualities that have made him such an outstanding ambassador of astronomy. Attention to detail; impeccable style even in foreign languages; felicitous references to the prides and achievements of the host country; pertinent understanding and analysis of the current trends of astronomical research; recognition and appreciation of the legitimate requirements of others - coupled with firmness and resolve to achieve mutual agreement. In international matters, as in personal matters, his integrity is unquestioned. In the course of his term as President (and both before and after) he represented organized astronomy on many formal, and informal, occasions - and the image of astronomy, and its truly international nature, has thereby been greatly enhanced.

If this brief tribute should have given the impression of a man whose contribution to international co-operation has been on a formal level, let it be corrected. Many are those who have enjoyed the warm hospitality, good fellowship and true friendliness of Mieke and Jan Oort.

When it could truthfully be said that one knew, if not personally then by correspondence or repute, all other astronomers, there was a general acceptance of the phenomenon that astronomers were specially "nice" people; there can be no doubt that personal contacts and friendships still play a large part in the furtherance of co-operation and, above all, of understanding. And Oort has many friends! One of my outstanding recollections is of an impromptu and alfresco "lunch" when (during the meeting of the IAU Executive Committee in Erevan, in 1962, which Oort attended as immediate past President), in a field overlooking Lake Sevan, some of the more senior of the world's astronomers ate freshly caught, and cooked, trout with nothing more sophisticated than bare hands. It was a most successful meeting!

In his 1961 Presidential address Oort, with justifiable pride, said that in all probability he was the only person to have attended every General Assembly of the IAU; this remarkable record has now been extended from 11 (Berkeley) to 16 (Grenoble). It is, however, insignificant in comparison to the contributions he has made, both within the Union and in so many other capacities, to the universality of the astronomical search for knowledge.

D.H. Sadler was with the Royal Greenwich Observatory until his retirement. He has been General Secretary of the I.A.U. from 1958-1964.

D.H. Sadler

REMINISCENCES OF THE EARLY NINETEEN-TWENTIES

Peter van de Kamp

I must have first met Oort on 27 May 1922, when Harlow Shapley gave a lecture in Leiden during his and Mrs. Shapley's visit to Holland. But my first clear recollection of meeting Oort refers to an event later that year. Oort, at that time assistant at the Kapteyn Astronomical Laboratory in Groningen, had been invited by Frank Schlesinger to come to the Yale Observatory, where he would be assistant over the interval 1922-1924. Oort's westward voyage had been preceded by two other Dutchmen: Adriaan van Maanen in 1911, and Willem J. Luyten in 1922, both of whom remained in the U.S.A. Kapteyn and van Rhijn had made temporary visits. "Countless" other young Dutchmen followed this westward trend: Peter van de Kamp (1923), Jan Schilt (1925), Dirk Reuyl (1927), Dirk Brouwer (1927), Bart J. Bok (1929), Gerrit P. Kuiper (1933), Gerard F.W. Mulders (1935) and many, many others. (I should like to include Kaj Aa. Strand, for many years Hertzsprung's assistant in Leiden.) While these and later young Dutch astronomers remained in the U.S.A., those who returned - and it was Holland's good fortune - included Jan Oort and Adriaan Blaauw.

In the summer of 1922 I was asked by Pieter Johannes van Rhijn, director of the Kapteyn Astronomical Laboratory (with the encouragement of Jan van der Bilt, lector in astronomy in Utrecht, with whom I had studied and worked for two years) to succeed Oort as assistant at Groningen. I clearly recall the interview with van Rhijn, and Oort's helpfulness. On 4 September 1922 I started my astronomical career in Groningen. How could I have anticipated that a few months later, at the invitation of Samuel Alfred Mitchell, director of the Leander McCormick Observatory at the University of Virginia in Charlottesville, I also would be on my way to the U.S.A.! This event was a direct result of the questions raised by Kapteyn (1922) in his last paper in B.A.N. no. 14 (Vol. 1, pp. 69-78), dealing with the problem of systematic errors in proper motions in declination. I travelled on the S.S. Rijndam, a miserable stormy voyage lasting 13 days, leaving Rotterdam on 20 February and arriving in New York on 6 March 1923, where I was met by amateur astronomer G. Ernest J. Yalden; this had been arranged by Harlow Shapley, at whose request I had brought a Moll-photometer.

Next I met Oort during the thirty-first meeting of the American Astronomical Society held 27-29 December 1923 at Vassar College,

H. van Woerden, W. N. Brouw, and H. C. van de Hulst (eds.), Oort and the Universe, 51–54.

CARNEGIE INSTITUTION OF WASHINGTON
MOUNT WILSON OBSERVATORY
PASADENA, CALIFORNIA

1923 May 7

Dear Dr. Slipher.

I hear from Russell, that you are looking round for some assistant. I therefore like to call your attention to two rather promising young astronomers from Holland, P. van de Kamp, who is for a year at the L. McCormick Obs. and Oort, who is at Yale.

Maybe either of these would be available before long.

Sincerely yours.

A. van Maanen.

Thanks [illegible]

[illegible]

Letter of recommendation about Oort to Slipher from van Maanen.
(Courtesy Dr. A. Hoag, Lowell Observatory)

Poughkeepsie, New York. It was a "large" meeting. Some fifty people attended and as "many" as twenty-three papers were presented, ranging from 5 to 10 minutes duration, several of them requiring "Lantern". Oort presented a paper on "The average radial velocity of double stars", Luyten on "Remarks on the luminosity curve". Apart from the few young people, the older, distinguished, astronomers were very much in evidence. They included: E.W. Brown, Benjamin Boss, Henry Norris Russell, Frank Schlesinger, John A. Miller, Frederic Slocum, Joel Stebbins, R.S. Dugan, J.Q. Stewart, Harlow Shapley and many others. The society photograph of 27 December shows, toward the rear, three young Dutchmen closely together: Luyten, Oort, van de Kamp. On Friday evening 28 December, a number of us, I dare say nearly all the "young" people, went to the local "Liberty" movie theatre, which presented a double feature (all of this for 15 cents), one of which was a "screen novelty" entitled A TRIP TO MARS. I have tried, in vain, to obtain a copy of this masterpiece, which may have contributed to the continuing astronomical career of our little group. My diary records that this show was attended by Oort, Luyten, Comrie and the following lady astronomers: Priscilla Fairfield (who in 1929 married Bart Bok), Cecilia Payne (who later married Sergei Gaposchkin), Adelaïde Ames and Lois Slocum.

My first serious interest in Oort's research began during those "golden days" 1918-1927, as Bertil Lindblad called them in his letter to me of 5 March 1965, written shortly before his death.

It really all started with Kapteyn's remarkable and admirable discovery (announced at the St. Louis Exposition Congress in 1904) of the "two star streams", or as it was called later, ellipsoidal distribution or preferential motion. Kapteyn's important discovery remained a mystery and defied explanation for more than two decades. But not any longer after the "galactocentric revolution", the breakthrough achieved by Harlow Shapley based on his study of the space distribution of globular clusters (1914-1917). Significant next steps toward understanding the state of motion of our Galactic System were the studies of the asymmetry in the motions of stars, globular clusters and galaxies by Gustav Strömberg (1924, Ap. J. 59, 228). This asymmetry consisted of an increase in group motion with increasing preferential motion, in the galactic direction of Argo (opposite Cygnus). Asymmetry had been recognized earlier for stars of large space motion, especially through studies by Oort (1922, B.A.N. 1, 133; no. 23). The time had now come to complete a synthesis of the galactocentric view of our Milky Way system and the kinematical properties of stars in our neighborhood. Bertil Lindblad (1926, several papers in Arkiv Mat. Astr. Fys. 19A, 19B and 20A) developed a successful model of galactic rotation which explained preferential motion and asymmetry. Lindblad's theory was confirmed through the discovery and study by Oort (1927, B.A.N. 3, p. 275; no. 120) of differential galactic rotation, the galactic equivalent of Kepler's third law. Thus the golden decade 1918-1927 witnessed the gradual falling into place of pieces which up to 1918 had been a puzzle, beginning with the "star streams".

Astronomers could now anew proceed with studies of the structure and dynamics of our Milky Way system. This is not the place to even give a brief indication of all that happened. I will end these recollections by stating how aware I am of the leading role which Oort has played and continues to play in these studies. Clearly Oort is the worthy successor to Kapteyn.

My own long-range work on unseen astrometric companions of nearby stars has always received recognition and continued encouragement from Oort. I should like to express my gratitude, or should I say, my pleasure, to fate that my astronomical life, through his work, correspondence and personal relationship, has been brightened by the presence of the greatest living astronomer: Jan Hendrik Oort.

Peter van de Kamp studied at Utrecht, and obtained his doctorate at Groningen in 1926. He worked at the Leander McCormick Observatory, Charlottesville and later became Director of Sproul Observatory, Swarthmore, Pennsylvania, and is now retired.

Peter van de Kamp.

The American Astronomical Society Meeting at Vassar College, Poughkeepsie, New York, 27 December 1923.

THE FIRST FIVE YEARS OF JAN OORT AT LEIDEN, 1924 TO 1929

Bart J. Bok

Gerard P. Kuiper, Pieter Th. Oosterhoff and myself, Bart J. Bok, descended on Leiden University and the Leidse Sterrewacht in late September, 1924. Willem de Sitter was then its Director and Ejnar Hertzsprung his chief associate. Jan Oort had recently been appointed a conservator (senior scientist) at Leiden. Oosterhoff had planned to study celestial mechanics under De Sitter and Jan Woltjer, but he was promptly taken in hand by Hertzsprung, who booked him permanently for work on variable stars. Kuiper turned to Hertzsprung and planned his own work on binary stars, the field of his ultimate doctoral thesis. During my high-school years in the Hague, I had developed an interest in our Milky Way system, inspired by the controversies of the early 1920's, which involved Harlow Shapley (whom I admired) and Heber D. Curtis - controversies in which Kapteyn and Easton figured prominently. Kapteyn had died two years before I arrived at Leiden. Small wonder that right at the start I turned to Jan Oort as my mentor. Jan and I share the same birthday, 28 April, but Jan is precisely six years ahead of me - a fact he never permits me to forget! Jan had come to Leiden as the former student of J.C. Kapteyn and Piet van Rhijn. To the new arrival Bart Bok, he was the man who had learned his trade from one of the great names of the recent past and from the man who had inherited Kapteyn's throne. It was a great thing to have come to Leiden, direct from a city high-school in the Hague. Woltjer and Oort were my guides during my first (three) years at Leiden. I took cues from van Rhijn and Oort during the two years that followed, and then my hero, Harlow Shapley, invited me to come to Harvard. To complete the picture, I should put down on paper that I arrived (by S.S. Veendam) in New York on 7 September, 1929 and that I married Priscilla Fairfield on 9 September, 1929.

Soon after I first came to Leiden, Jan Oort finished drafting his dissertation on High-Velocity Stars, a magnificent piece of research that was published in the famous Groningen Publications, no. 40 (1926). His doctorate and his marriage to Mieke followed in quick order.

Jan had a wellpublicized argument with Gustaf Strömberg of Mount Wilson Observatory (see Observatory, Vol. 49, pages 279 and 302, 1926), in which the two of them argued especially about the sharpness of the cut-off at 63 km sec^{-1} in the asymmetry of directions of motion of

H. van Woerden, W. N. Brouw, and H. C. van de Hulst (eds.), Oort and the Universe, 55–58.

high-velocity stars. Strömberg had a vague theory of explaining asymmetry in terms of relativistic restrictions in directions of motion (which did not seem to make sense at the time when it was first proposed), but Jan Oort was seeking a solution in terms of the new galactic rotation theory, then being developed by Bertil Lindblad.

To the best of my knowledge, the mathematical approach to the theory of galactic rotation originated with Bertil Lindblad, who - in his early papers on the subject - approached it primarily from a purely mathematical point of view. He was interested in explaining two phenomena in terms of a rotating galaxy. First on his list was the phenomenon of the high-velocity stars and the marked asymmetry in the distribution of their directions of motion in the galactic plane. Second came the interpretation of the phenomenon of the ellipsoidal distribution of stellar motions. The ellipsoidal theory had by the 1920's pretty well replaced Kapteyn's earlier interpretation in terms of two star streams. Lindblad's treatment was primarily a mathematical one. I remember how, as young students, Kuiper and I discussed that we obviously needed to know a lot more differential geometry (Bertil Lindblad was always referring to "osculating planes" and the like) before we could hope to understand how to interpret properly asymmetry in stellar motions and ellipsoidal distributions. Anyhow, Jan Oort was presenting (on Monday afternoon at four) a series of seminars for Doorn, Kuiper, Oosterhoff and Bok on Lindblad's theories of galactic rotation. As I remember it - others may have different recollections! -, Jan told the four of us one Monday that he had got bogged down in Lindblad's complex mathematics and that there would be no lecture the next Monday afternoon. And, as I remember it, there were no lectures for two Monday afternoons to follow. And then there came the first Monday lecture after the crisis, a lecture in which Jan Oort basically developed the simple formulae for the double sine-wave effect of galactic rotation in radial velocities and the corresponding formulae for the effects in proper motions. The four of us realised that we were listening to an amazing new step in the understanding and interpretation of stellar motions, a step that Bertil Lindblad had overlooked in his differential geometry oriented approach to the subject! Jan gave us the simple formulae simply derived, and we left the lecture room thoroughly aware of the fact that the effect of galactic rotation on radial velocities is proportional to the average distance of the group of stars under consideration. O and B stars and long-period cepheids were listed by Jan Oort as the key objects for future studies.

We learned one other thing on that first afternoon: the effects on proper motions in galactic longitude were highly dependent on precessional corrections, and very much affected by possible systematic errors in the fundamental system of proper motions - as they are today. Oort's constant A seemed determinable; B was a different matter.

Publication of Jan Oort's work in B.A.N. 120 (followed soon by B.A.N. 132) and a paper in the proceedings of the Royal Dutch Academy of Sciences came next, or at about the same time. By 1927 all of us interested in understanding stellar motions were made to feel at ease, for the great mysteries of the early and middle twenties had been resolved.

Bok, Oort, van der Kamp and Schilt at the Harvard Summerschool, 27 July 1937.

There followed some wonderful years: 1928 was the year when the IAU met in Leiden. J.S. Plaskett came to Leiden on crutches after a severe illness, to talk with Oort and Lindblad about his plans for radial-velocity work on O and B stars to check firmly and once for all the basic tenets of the double-sinewave approach to differential galactic rotation. He and J.A. Pearce proved their point.

Jan Oort rounded out his great contributions to the development of galactic rotation with a magnificent paper on the dynamics of our rotating galaxy, published in B.A.N. 159. I keep a copy of B.A.N. 159 handy for reference when I am now asked to give a lecture on the dynamics of our rotating galaxy. The paper is still a fine basic reference fifty years after it was first published.

During my final years in Holland, Jan Oort turned his attention to a field of research that had been dear to Kapteyn's heart toward the end of his life - the relations between stellar distributions and stellar motions perpendicular to the galactic plane. Another classic Jan Oort paper followed: B.A.N. 238. It was indeed a privilege for me to have become associated with Jan Oort during his period of remarkable early productivity. I had a ring-side seat in what was the greatest show ever in stellar motions. After galactic rotation was established, there remained some mopping up to be done. R.J. Trumpler solved the most vexing problem: Is there any general absorption of light close to the galactic plane? The affirmative answer came in 1930.

Under the continued leadership of Jan Oort, our knowledge of the structure, kinematics and dynamics of our Galaxy has continued to develop. His collaboration and friendship with Walter Baade helped pave the way toward the discovery of spiral structure of our Galaxy by Morgan, Osterbrock and Sharpless. Oort's encouragement of the work by Henk van de Hulst, together with Jan Oort's emphasis on radio astronomy, helped bring about the detection of the 21-centimeter line of neutral atomic hydrogen by E.M. Purcell and H.I. Ewen. It was Jan Oort who, in the middle 1950's, led the Dutch assault on the radio spiral structure of our Galaxy. C.C. Lin has told all in the field that the first suggestion for the successful density-wave theory of galactic spiral structure came from Jan Oort at an early 1960's symposium at Princeton. Jan Oort has indeed left his mark on the study of our Milky Way galaxy. The Milky Way has not been the same since Jan Oort undertook to look at it in earnest in the early 1920's. Jan was the great and worthy successor to Kapteyn.

Bart J. Bok studied at Leiden (1924-29), and obtained his doctorate at Groningen in 1932. He worked at the Harvard, Mount Stromlo and Steward Observatories, and is now retired.

BART J. BOK

EARLY GALACTIC STRUCTURE

Per Olof Lindblad

In the course of the 1920's the true nature of the stellar system, its general state of motion, and its relation to the spiral nebulae became understood.

In the beginning of the decade a considerable amount of observational data concerning the spatial arrangement and motions of stars had been obtained. The extensive work by Kapteyn and van Rhijn had enabled them to suggest a model for the entire observable stellar system (Kapteyn 1922). The sun appeared to be at the centre of the system and in the Kapteyn-van Rhijn model the equidensity surfaces would be similar ellipsoids of revolution with axial ratio 5:1, the shortest axis being perpendicular to the galactic plane. In the plane the density would have dropped to 40% at a distance of 1 kpc from the centre and to 10% at 2.8 kpc. Kapteyn estimated that the true maximum distance of the sun from the centre of the system must be less than 700 pc. From the number density of stars near the sun a total number of 5×10^{10} stars was derived. Indeed, the existence and influence of dark matter was discussed at the time but its properties and galactic distribution were not fully appreciated.

In contrast to these results Shapley (1918) had found that the globular clusters, their estimated distances ranging from 6.5 to 67 kpc, formed a system of vastly larger volume than the Kapteyn system, but with a plane of symmetry definitely coinciding with the galactic plane. It occupied essentially only one hemisphere and its centre would fall in the direction of Sagittarius ($\ell_{II} = -2^{\circ}$) at a distance estimated by Shapley at about 20 kpc.

Already in 1904 Kapteyn had found that the proper motions of stars indicated the existence of two interpenetrating star streams. The relative velocities of those streams would be 40 km/s in a direction parallel to the galactic plane along the line $\ell_{II} = 20^{\circ} - 200^{\circ}$. K. Schwarzschild in 1908 showed that this star streaming could be represented by an ellipsoidal frequency function with the longer axis in the direction of Kapteyn's star streaming.

Radial velocities, however, revealed an asymmetry of the larger motions that seemed to be preferentially directed towards one hemisphere. Strömberg (1924) at Mt. Wilson analysing space motions studied this phenomenon. He divided the material according to different classes of

H. van Woerden, W. N. Brouw, and H. C. van de Hulst (eds.), Oort and the Universe, 59–64.

objects and found a linear relation between the group motion and the square of the velocity dispersion. The group motions were directed parallel to the galactic plane almost at right angles to the direction of star streaming, and amounted in the extreme cases to 260 km/s for the globular clusters and 390 km/s for the non-galactic nebulae. As Strömberg points out, the direction of the asymmetric drift motion also is almost at right angles to the direction of the centre of Shapley's system of globular clusters. He suggests as one of two possibilities that the local Kapteyn system moves around this distant centre with a velocity of 300 km/s.

On the theoretical side Eddington and Jeans had developed the dynamical theory for stellar systems in equilibrium. Introducing two integrals of motion, Jeans' (1922) solution admitted drift motion in the tangential velocity coordinate. Jeans applied this to the phenomenon of star streaming and the Kapteyn stellar system assuming, as did also Kapteyn, that star streaming was caused by stars orbiting around the Kapteyn system in opposite directions.

In his thesis of 1926, printed as Publication from the Kapteyn Astronomical Laboratory in Groningen No. 40, Jan Oort presents a detailed study of the high-velocity stars. Already in 1922 Oort had pointed out that the asymmetry of large motions becomes obvious above a critical value of about 63 km/s. Now Oort further strengthens this conclusion and stresses its importance. Stars above this velocity have their space motions almost entirely directed within the longitude interval $\ell_{II} = 170^\circ$ to 10°.

Oort points out that the planetary nebulae show a concentration towards the same direction as do the globular clusters, and presents the arguments why the globular clusters are connected with the system of high-velocity stars in the Milky Way. He considers the core of the Kapteyn system to be a local concentration within this larger system, which comprises the high-velocity stars and globular clusters, and moving relative to its centre. His suggestion is that the critical velocity of 63 km/s is the velocity of escape from the local system into the larger one. Oort showed that the frequency maxima of the directions of high velocity for all but the very highest fell close to the ends of the longitude interval occupied, in contradiction to what could be expected in the case of a large velocity of the sun with respect to a system with high isotropic velocity dispersion. Oort suggests that star streaming among the high-velocity stars could explain this particular phenomenon, but rejects the idea on account of the distribution of the very highest velocities.

In the meantime B. Lindblad (1925) applied to the big stellar system an analysis related to the one applied by Jeans to the local Kapteyn system. He assumed that the big stellar system could be divided into a number of sub-systems with rotational symmetry around the same axis but with different mean velocity of rotation and different degree of flattening. The most flattened system with the highest speed of rotation was identified with that of the Milky Way star clouds, while the least flattened with the slowest rotation was identified with the system of globular clusters. With the sun moving with the Milky Way clouds, the slower rotating systems would show an asymmetric drift in a direction at right angles to that of the centre.

Solving the differential equations given by Jeans for the velocity dispersion in a stellar system in dynamical equilibrium, Lindblad derived a quadratic relation between velocity dispersion and mean rotational velocity, which could be compared with the relation derived from observations by Strömberg. Zero velocity dispersion giving maximum rotational velocity would correspond to circular orbits. Assuming this maximum velocity to be 350 km/s and the distance to the galactic centre 12 kpc, a total mass of the Galaxy of 1.8×10^{11} $M_{\odot}$ was derived. Lindblad (1926a) identified Oort's critical velocity of 63 km/s with the relatively small difference between the circular velocity and the velocity of escape close to the edge of the system. At the time he considered the total avoidance of one hemisphere, shown by the large velocities, to be the best evidence for the correctness of his theory (1926b).

Lindblad had assumed that all the sub-systems had the same extension in the galactic plane. Within each sub-system the density might vary with distance from the plane, but it was assumed that the surfaces of constant density were spheroidal surfaces with the same diameter in the galactic plane. Then, in his system each sub-system would rotate with constant mean angular velocity in the galactic plane and there would be no differential rotation in the direction of the radius for a particular sub-system. Also, in such a system the velocity dispersion parallel to the galactic plane will be the same in all directions, which fact for some time concealed the true cause of Kapteyn's star-streaming.

After some original scepticism Oort became very positive to Lindblad's hypothesis. He realised that, as in the general case the rotation would not be one of constant angular velocity for the various sub-systems, effects of differential rotation would occur that might be observed.

In April 1927 Oort sends Lindblad a letter with a proof of his paper on the differential rotation as shown by the radial velocities. I know that this letter made my father particularly happy, not only because of the support Oort's paper gave to his hypothesis, but this was also the first sign of appreciation for his idea that he received from an outside colleague.

The paper Oort enclosed with the letter had the title "Observational evidence confirming Lindblad's hypothesis of a rotation of the galactic system" (1927a). In his analysis of the effects on proper motions and radial velocities Oort here introduced the well-known constants of differential rotation A and B; r being the distance from the sun, rA is the amplitude of the variation of radial and tangential velocities with galactic longitude and B a constant term in the proper motion. Together they give the angular velocity of rotation and its variation with distance from the galactic centre in the solar neighbourhood. Oort found that indeed there was a periodic variation of the radial velocity with galactic longitude and a phase of the variation such that the direction to the centre of rotation agreed with the direction to the centre of the system of globular clusters. Different types of objects with different amplitudes in the velocity variation, with the exception of O and M stars, gave very consistent values of A.

In return, the amplitudes being proportional to the distance, Oort could estimate mean distances of objects of otherwise unknown distances like planetary nebulae and interstellar calcium clouds.

In order to determine the constant term B in the proper motions, Oort had to determine an absolute value of the precessional constant from proper motions in galactic latitude (1927b).

Assuming circular motions, A and B would give the linear part of the force law in the galactic plane, and Oort from these values estimated the relative contributions of a galactic disk and a central bulge.

The angular velocity being known, an estimate of the corresponding velocity in a rest frame would give the distance to the galactic centre. Assuming that the system of globular clusters possessed no rotation, Oort found the mean rotational velocity of the observed stars to be 272 km/s and derived a distance to the galactic centre of 6 kpc.

Such a distance made the centre of the Galaxy fall far outside the effective limit of the Kapteyn universe. Thus there must exist a large mass of unobserved matter. Oort concluded that the discrepancy was due to obscuration by dark matter in the galactic plane.

The month after receiving Oort's letter with the preprint of his paper, Lindblad (1927) submits a paper explaining the cause of the ellipsoidal distribution of stellar velocities, giving the ratios of the dispersion axes in the galactic plane as a function of Oort's constants A and B. Lindblad here introduces the concept of epicyclic motion into galactic dynamics and shows that a gaussian distribution of sizes of epicyclic orbits leads to an elliptical velocity distribution of velocity components parallel to the galactic plane, the ellipse being elongated towards the galactic centre.

In a second classical paper with the title "Dynamics of the galactic system in the vicinity of the sun" (1928) Oort presents a very fundamental analysis. It is based on the studies by Eddington and Jeans, and the starting point is the collision-less Boltzmann equation for the case of a stationary distribution function for the stars in phase space. Introducing for the distribution function a general Schwarzschild ellipsoid, admitting a drift motion in the direction of rotation and inserting this in the Boltzmann equation, Oort derives a number of equations of condition.

From these Oort can show that the velocity dispersion in the radial direction is independent of the distance to the centre, while the ratio of the axes of dispersion in the radial and tangential directions again shows the same relation to the constants A and B as derived in a quite different way by Lindblad. The definitions of A and B have here been generalised to the case of systems with a mean rotation different from that of circular motion.

Further, Oort derives the quadratic relation between asymmetric drift motion and velocity dispersion. In his derivation of the corresponding relation, Lindblad had assumed ellipsoidal space density distributions with a common limiting radius for the different sub-systems in the galactic plane. Then the parameters of the relation came to contain the ratio of the sun's distance from the centre to the limiting radius. In Oort's relation, on the contrary, the parameters in a more natural way contain the stellar density gradient of the sub-system considered. Oort shows that comparison with observation gives a rather consistent density gradient. Its estimated magnitude for the stars treated is such that the density would have increased by a factor of 2 at a distance 1 kpc closer to the centre.

Thus, in a very beautiful way Oort had derived the fundamental kinematic properties of stars and related them to the total distribution of matter in our Galaxy as well as to the distribution of the particular subgroup of stars considered.

Unexplained still was the observed vertex deviation of the velocity ellipsoids. Both Oort and Lindblad attributed this phenomenon to the presumed spiral structure of the Galaxy. Their rather different explanations have been summarized by Oort (1939, 1940).

There remained to be exploited the motions perpendicular to the galactic plane. Improving the first attempt by Kapteyn (1922) Oort in a fundamental study (1932) takes up the investigation of the force by the stellar system in the direction perpendicular to the galactic plane and the stellar density distribution in this direction. If one assumes the motions of stars at right angles to the plane to be uncoupled from the motions parallel to the plane, the density distribution will be a function of the energy integral and thus a function of the velocity and gravitational potential. Discussing available observational material Oort derives the force law out to $|z| = 600$ pc, the equidensity surfaces for stars in the solar neighbourhood which indicate an increase of density towards the centre of the Galaxy, and finally the total density of matter near the sun. Later Oort (1960) improves the method searching a solution which also satisfies Poisson's equation.

Bertil Lindblad and Jan H. Oort.

When radio waves from the celestial sky were detected, Oort was one of the first astronomers to realise the importance of this discovery and in particular the importance for galactic studies if lines in the radio spectrum could be detected. With van de Hulst, Muller and other collaborators he after the war worked on the Dutch part in the detection of the 21-cm line of interstellar atomic hydrogen and on the exploitation of this discovery in galactic research. At the General Assembly of the International Astronomical Union in Rome in 1952 Oort could present the first results of this research. Here the spiral structure of the outer part of the Galaxy and in particular what is called the Perseus arm, was first discerned on a general scale (van de Hulst, Muller and Oort, 1954). The radio line studies thus introduced meant a new revolution in galactic research.

References

Jeans, J.H. 1922, Monthly Notices Roy. Astron. Soc. 82, pp. 122.
Kapteyn, J.C. 1922, Astrophys. J. 55, pp. 302.
Lindblad, B. 1925, Arkiv mat. astron. fysik 19A, No. 21 (= Uppsala Medd. No. 3).
Lindblad, B. 1926a, Arkiv mat. astron. fysik 19A, No. 35 (= Uppsala Medd. No. 13).
Lindblad, B. 1926b, Populär Astronomisk Tidskrift 7, pp. 125.
Lindblad, B. 1927, Arkiv mat. astron. fysik 20A, No. 17 (= Uppsala Medd. No. 26).
Oort, J.H. 1922, Bull. Astron. Inst. Netherlands 1, pp. 133 (No. 23).
Oort, J.H. 1926, Publ. Kapteyn Astron. Laboratory Groningen No. 40.
Oort, J.H. 1927a, Bull. Astron. Inst. Netherlands 3, pp. 275 (No. 120).
Oort, J.H. 1927b, Bull. Astron. Inst. Netherlands 4, pp. 79 (No. 132).
Oort, J.H. 1928, Bull. Astron. Inst. Netherlands 4, pp. 269 (No. 159).
Oort, J.H. 1932, Bull. Astron. Inst. Netherlands 6, pp. 249 (No. 238).
Oort, J.H. 1939, Monthly Notices Roy. Astron. Soc. 99, pp. 369.
Oort, J.H. 1940, Astrophys J. 91, pp. 273.
Oort, J.H. 1960, Bull. Astron. Inst. Netherlands 15, pp. 45 (No. 494).
Shapley, H. 1918, Astrophys. J. 48, pp. 154.
Strömberg, G. 1924, Astrophys. J. 59, pp. 228.
Van de Hulst, H.C., Muller, C.A., Oort, J.H. 1954, Bull. Astron. Inst. Netherlands 12, pp. 117 (No. 452).

Per Olof Lindblad, son of Bertil Lindblad, is at the Stockholm Observatory and at European Southern Observatory, Geneva.

EARLY GALACTIC RADIO ASTRONOMY AT KOOTWIJK

C.A. Muller

On a snowy day in December 1950 I bicycled through the woods from Apeldoorn to Kootwijk-Radio, the central transmitting station of the Dutch Post Office, to start my work for the Netherlands Foundation for Radio Astronomy. Some months before I had finished my studies in physical engineering at the Delft Technical University and this was my first job. I was to continue the work started in 1948 by Mr. Hoo towards the discovery of the 21.2 cm line of neutral hydrogen. This line had been predicted by H.C. van de Hulst, then still a student, in a paper on the origin of the radio waves observed by Jansky and Reber. It was presented at a colloquium on "Radio Waves from Space" held in 1944, during the war, at Leiden Observatory.

This colloquium marks the beginning of Dutch galactic radio astronomy. It was probably the first time that professional astronomers discussed the possibilities of radio astronomy and it is obvious that Oort had stimulated this meeting. If the hydrogen line could be observed, then there were exciting new possibilities for the study of the structure of our Galaxy. But a large radio telescope of some 25 meters would be necessary and as early as 1945, a few months after the end of the war, Oort made a preliminary proposal for such an instrument to the Dutch Academy of Sciences. However at that time this plan could not be realized, though it met with great interest. In the next few years radio astronomy in Australia and England led to many fascinating discoveries which strongly stimulated the interest in radio astronomy. Both the Utrecht Solar Observatory under Professor M. Minnaert and the Radio Department of the Post Office became interested in solar observations and started experiments on a limited scale. The common interest in radio astronomy by the Leiden and Utrecht Observatories, by the Post Office and by scientists at the Philips Physical Laboratories at Eindhoven, led in 1949 to the founding of the Netherlands Foundation for Radio Astronomy, with Oort as the first chairman. At the Post Office it was Mr. A.H. de Voogt, one of the chief engineers of the Radio Department, who had an interest in solar radio astronomy because of the relationship between sun and ionosphere. He had rescued a few of the $7\frac{1}{2}$-meter Würzburg antennas, which had been part of a German radar chain along the coast during the war, from destruction and had them repaired

H. van Woerden, W. N. Brouw, and H. C. van de Hulst (eds.), Oort and the Universe, 65–70.

for radio-astronomical purposes. One of these antennas was made available to the new Foundation for its hydrogen-line experiments. It had been placed on the southern slope of a small hill at the Kootwijk-Radio transmitting station, overlooking a beautiful area with heath and woods, but uncomfortably close to high-power transmitting antennas!

Here I started my work with one assistant. Though the experiments had been going on for two years, I had to start almost from scratch because all receiver equipment had been destroyed in a small fire earlier that year. However some parts for a new receiver were already under construction at the Philips laboratories under the supervision of Mr. F.L. Stumpers. I knew almost nothing of astronomy or radio astronomy at the time, and looking back it seems a small miracle that some five months later we observed the 21-cm line. I think this miracle was possible because all the conditions were favourable for it. The discovery of the line was primarily a technical problem of constructing a suitable receiver, and it was a great help that I was working in the almost ideal surroundings of the small transmitter-construction division at Kootwijk with its group of enthousiastic collaborators, a well-equipped laboratory and a large workshop, which worked quickly and efficiently. With my experience as a fervent radio amateur I fitted quite well into this group. The same group also had experience with equipment for radio astronomy because it had constructed the receivers for solar observations.

Our discovery of the line was hastened by the fact that Van de Hulst was visiting the United States at the time. At Harvard he met H.I. Ewen, who a few weeks later would make the first observations of the 21-cm line. In a short letter which I received early in March, he told me that Ewen was working for his thesis on a hydrogen-line receiver with frequency-switching in the second local oscillator. This information came just at the right moment. Our first experiments with part of the receiver had clearly shown that the stability of a simple non-switching receiver would be insufficient, and from the literature I had available, it was clear that it would be necessary to use some form of Dicke receiver in which the input was switched periodically beteen the antenna signal and the signal from a noise source. A synchronous detector would then measure the difference between the two signals. Then the letter arrived with the proper solution to our problem: the noise-source signal could be replaced by an antenna signal at a different frequency. The construction of the frequency-switched receiver did not take too much time. We modified the Harvard concept somewhat by using frequency-switching in the first local oscillator, by adding a reactance frequency modulator to the 6.4-MHz oscillator of the crystal-controlled Philips-built frequency multiplier. We also added a tunable second local oscillator and a narrow-band second i.f. amplifier as well as a 30-Hz amplifier-synchronous detector section and some calibration facilities, and then we were ready for our first attempts to observe the line. I think it was on the second night, on May 11, 1951, that the line was found.

I remember very little of that night. There is a vague recollection of sitting in the telescope cabin on a nice spring evening, switching the second local oscillator every few minutes between the two frequencies, which in the presence of the line should give alternately a positive and a negative deflection on the recording meter, while a region near the galactic plane drifted through the antenna beam, but that is all. I think that at that time I hardly realised the importance of what I was doing and perhaps I had enough confidence in our equipment to expect the line to show up. At that time I must have known that it was there, because Ewen had already observed it some six weeks earlier.

Then for a month we tried to make a series of observations near the galactic plane, but though the line was clearly observed, the reproducibility of these observations was poor and insufficient for systematic measurements of the hydrogen distribution. One of the best observations appears below. It shows a drift curve through the Cygnus region at a fixed frequency, with the characteristic alternating positive and negative deflections due to the switching of the second local oscillator frequency at intervals of a few minutes.

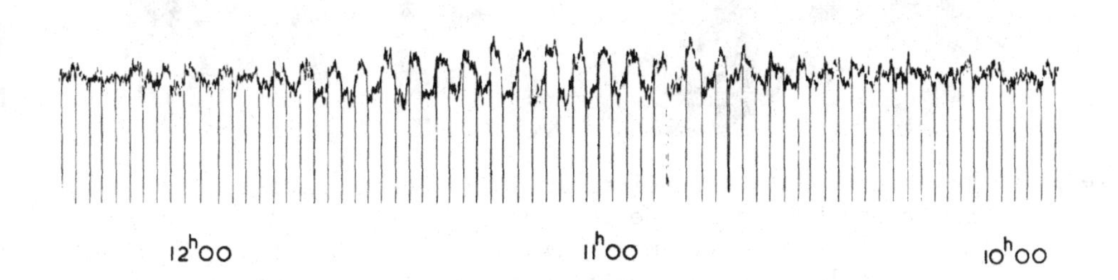

Figure 1 One of the first hydrogen line observations. It shows a drift curve through the Cygnus region.

The more or less simultaneous discovery of the 21-cm line in the United States, in the Netherlands and also in Australia, where W.N. Christiansen and J.V. Hindman observed it in June or July, raised great interest in the scientific world. In the Netherlands Oort as chairman of the Foundation for Radio Astronomy now managed to obtain a grant for a 25-meter radio telescope, the later Dwingeloo Radio telescope, which was put into operation in 1956. Another Dutchman, B. van der Pol, then director of C.C.I.R., took the first steps to protect the hydrogen line frequency from interference from other radio services.

It was not until a year later, in June 1952, that at Kootwijk we started the first hydrogen-line survey along the galactic plane with almost new equipment, helped by many students and staff members of the observatories who did the actual observations. Observing was not an easy job in this pre-automation era. During each two-hour observation of a line profile the telescope had to be moved by hand every few minutes, using two handles in a corner of the telescope cabin and reading

Figure 2 The hydrogen line receiver in 1951. In the middle the crystal mixer and behind it the frequency multiplier. At the right a box with the i.f. amplifier section and in front of it the 30 Hz amplifier and synchronous detector section. At the left power supplies and control equipment. Receiver and antenna are connected with a coaxial cable through the back wall of the cabin. The recording meter is not shown. The coaxial 1420 MHz amplifier lying in the middle next to the mixer was not used.

the azimuth and declination scales through small windows in the floor and a side-wall of the cabin. In addition there were several receiver functions which had to be watched and adjusted, including one in a box at the back of the $7\frac{1}{2}$-meter reflector. The first survey lasted one year, then the receiver was revised and extended. The second survey, which covered a strip along the galactic plane with a width of 20° in latitude, started in November 1953 and lasted until August 1955, when our small group, then consisting of four persons, left for Dwingeloo. These two surveys led to the well-known maps of the large-scale structure of the part of the Galaxy that is visible from the Netherlands.

The year between the discovery of the hydrogen line and the beginning of the first survey I remember as a most productive period in my life. It meant understanding and solving all problems of a new technique. We learned how to build suitable components for a radio astronomy receiver and we also learned to cope with strong interference from the nearby transmitters. It laid the foundation for further receiver developments in later years, in which a steady improvement in sensitivity and stability took place as well as a gradual development towards multi-channel receivers. It was not until many years later that I realized how difficult the long delay between the discovery and our first systematic observations must have been for Jan Oort. I still appreciate very much the fact that he never showed any impatience or annoyance, but just had confidence in us and gave his full support. This attitude towards us in those early days and in fact during all the years I worked with him in radio astronomy is, I think, typical of his attitude towards the instrumentalist in his field of science. It is obvious that our work then and later was to a large extent just technical support for radio astronomy. It is not unusual that a scientist considers to some extent his own scientific activities as superior to the activities of those who supply him with the essential instrumentation or in some other way support his work. Jan Oort has always given me the feeling, not just in my own relationship with him, but also by the way he talked about the work of others, that for him these distinctions do not exist. It is not so much the type of work you are doing that matters, but the way you are doing it. It is this quality and his friendly interest in one's work that has made cooperation with him so valuable for almost twenty years.

C.A. Muller has been with the Netherlands Foundation for Radio Astronomy from 1950-1971. He is now Professor of Electrical Engineering at the Twenthe University in Enschede.

The 25-meter dish at Dwingeloo, as modified in December 1973.

OORT AND HIS LARGE RADIOTELESCOPE

W.N. Christiansen

What does one say about that period in Netherlands radioastronomy which I am trying to describe, 1960-1964 ? Now in 1980 we can look at it as the time of the birth pangs of the Westerbork Synthesis Radio Telescope, but at the end of 1963 how did it look ? Some might have described it as the time of the frustration of the hopes and plans of Jan Oort for the Benelux Cross Antenna; for that telescope was never built.

But does one think of Jan Oort in terms of frustration or failure ? The rejection of the Benelux Cross showed again his capacity, by thought and determination, for turning what seemed to be failure into triumph.

I was fortunate to be associated with Jan Oort, the man and the astronomer, during most of the planning of the Benelux Cross. I learned a great deal about the method of work that during the last half century has made his contribution to astronomy so great. I also found a great friend who has taught me to love a country so far removed in distance and culture from my own.

It is typical of what perhaps one may be permitted to call the inspirational nature of Jan Oort's approach to science that he was one of the very few professional astronomers to greet enthusiastically the early discoveries of radioastronomers. Doubtless his enthusiasm was based on the fact that the new technique gave for the first time the opportunity to study those parts of our Galaxy not visible to optical astronomers. In addition, one can surmise that he would not have been indifferent to the fact that in this branch of astronomy the frequently cloudy skies of his beloved country would no longer be a barrier to observational astronomy.

During the 1950's the pioneering work in the Netherlands and the United States on the 21-cm hydrogen line and the mapping of neutral atomic hydrogen in our Galaxy was one of the most exciting things that had happened in astronomy in many years. My first contact with Jan Oort was in 1951 when I received a letter from him asking for a copy of a rather crude map of peak 21-cm H-line emission from our Galaxy that Hindman and I had just completed. We felt very honoured to be recognized by the great Professor Oort !

H. van Woerden, W. N. Brouw, and H. C. van de Hulst (eds.), Oort and the Universe, 71–78.

It was apparent that the resolving power of existing telescopes was inadequate to trace the structure of the Galaxy and Jan Oort set about building the Dwingeloo 25-metre radio telescope, which for a short time was the world's largest paraboloid. Meanwhile, a number of radio-telescopes of novel design had been invented and built, mainly in Australia and in England. These had resolving powers which, although very poor compared with any optical telescope, were very much superior to radio telescopes of classical design.

One of these new radiotelescopes was the Mills Cross Antenna. Charles Seeger at Leiden proposed to Oort that an extremely large telescope of this type should be built in the Netherlands. Oort realized that it would be very difficult to obtain the very large funds required to build this telescope in the Netherlands, but thought that it might be possible to make it a project of the newly-formed Benelux group of countries. He persuaded the Netherlands governmental authorities to approach their Belgian counterparts with the suggestion that a Benelux radiotelescope should be built. The result of this was that a committee, with Oort as chairman, was set up. Funds for a design study of a Benelux Cross telescope were promised by the Netherlands and Belgian governments and by the OECD.

The first report to this committee was submitted by van de Hulst and Seeger. It specified a Cross Antenna to operate at 408 MHz, with a resolving power of one arcminute and a collecting area of 600,000 m^2. Such an instrument would be capable of making a number of observations that Oort considered to be important. These included observations of weak radio sources and an accurate determination of their position for optical identification, a study of the structure of individual radio sources and possibly also of normal galaxies. In addition a resolution of one arc minute could allow spiral structures of about 17 nearby galaxies to be resolved.

Oort realized that a wavelength of 21 cm would be much more desirable for the operation of the telescope than the chosen 74 cm wavelength, but it was felt that with existing receiving equipment the sensitivity would be inadequate at the shorter wavelength.

The first task of the Benelux Cross Committee, after it had accepted the specification just quoted, was to build up a group capable of designing and building such a telescope if the necessary funds could be obtained. Oort looked for someone who had had experience in building large telescopes and asked me to come to Leiden for a year or two. I had designed two radiotelescopes, which at the time had the highest resolving power of any that had been built. The newer one was a grating-cross telescope of 3 arcmin resolving power at a wavelength of 21 cm and was directly relevant to the Benelux project. The older telescope had been used as an earth-rotation synthesis telescope in the early 1950's, but the very long computing time (by hand) to make a single observation had led to its replacement by the grating-cross telescope. It is ironical that the progress of the Benelux project was to be in the opposite direction to that in Australia.

The early design group assembled by Oort consisted of Hooghoudt who was in charge of the mechanical design and Högbom who had taken his doctorate with Martin Ryle at Cambridge. I was to replace Seeger who

Leiden Observatory with director's house on the right.

intended to return to the United States. Two of Oort's own astronomical group, Muller and Westerhout, were to advise the design group when necessary. The group gradually increased in size, with foreigners outnumbering the Dutch and Belgians during most of the time. The first report on a design which accorded with the Benelux Committee's specification was prepared very rapidly. Two designs were submitted; one made use of cylindrical parabolic reflectors with stretched-wire reflecting surfaces; the other was an array of parabolic dishes. After studying the report, the Benelux Committee decided that the first design was preferable and the design group went ahead on more detailed design work.

At this stage Oort, who was still uncertain about the specified wavelength of operation, arranged for an international conference of people who were interested in high-resolution radio observations to be held in Paris by the OECD. The Conference was useful and there was strong support for the building of telescopes with resolving power of one or two arc minutes. The only new projects that were discussed were the Benelux Cross and the Bologna Cross telescopes. Although it was not mentioned at the Paris Conference, plans for a new telescope at Cambridge, England were disclosed in the press a few weeks later. Few details were given but it appeared that the new telescope would use the earth's rotation to synthesize a two dimensional aperture as had been done with the early Australian grating telescope. However, instead of an array of many paraboloids the new telescope would use only three paraboloids, one of which would be movable so that a complete observation would require many days.

The way in which the new telescope had been announced (or not announced) made it obvious that there might be rivalry in the building of a large radio telescope. Oort has never believed in races. "If people try to get results quickly, the work is probably not very good" I heard him say on one occasion, when I enquired about the delay in publishing some work which was being repeated elsewhere. The Cambridge design was considered to be too slow in operation by Oort and there was no tendency to turn the Benelux Cross into a synthesis instrument.

During the following months the Benelux Cross design group increased in size and the detailed design and site selection were beginning to reach finality.

The work was done at the Sterrewacht in Leiden. The foreigners, in the majority in the group, were fascinated by the atmosphere in the old Sterrewacht, now alas vacated by astronomers. In the 1960's salaries and prices in Holland were very low (this was a problem in working with the Belgians, where salaries and prices were high). The automobile had not spread to the plague proportions that it has now; most people came to work on bicycle and Leiden was a quiet place. To the foreigners life seemed rather old-fashioned and delightfully comfortable; to local people on low wages the second of these was not quite so evident.

Jan Oort at the Sterrewacht, in his office and in that very special house always open to visitors, both Dutch and foreign, was completely in the tradition of the great Dutch scientists and painters. Each morning when all the people at the Sterrewacht gathered for coffee, Oort would move slowly around the rooms, his notebook in his hand, questioning one after another, answering questions and writing down brief notes. A large part of his administration thus took place during morning coffee, leaving more time for "work". At morning-coffee time the large Australian contingent usually made its presence felt by an increase in the noise level. At the time the young Dutch astronomers were rather serious and sober in demeanour, with a profound dislike of attempting anything without a thorough knowledge of the subject and without long contemplation. It is doubtful if they have ever recovered from the influx of free and easy characters from the other side of the world, willing to tackle anything, even on the basis of minuscule knowledge. Even the Australian language spread, sometimes in an embarrassing way. I remember on one occasion, at a solemn meeting of the Benelux Committee, a Dutch engineer quoting my opinion verbatim, in raw Australian terms not at all suitable for printing here.

If Australian attitudes had some effect on the Dutch astronomers, the Sterrewacht and Jan Oort had a much greater influence on the Australians and other foreigners. I well remember one winter's evening near midnight when I was working on a report for the Benelux Committee. There was a knock on the door and Jan Oort's voice could be heard from outside "There is a full moon tonight, let us go for a walk Chris". We drove out of Leiden and then walked along a canal for some miles, through farms with the moon shining on the snow-covered roofs and above them lighting the edges of the beautiful clouds that are such a feature of the Netherlands. The paintings of the old Dutch landscape artists suddenly became alive in my mind and from that time Holland has had a very special place in my heart. Jan Oort's love of his country is so intense that one has to share it.

There are so many memories that I have of life and incidents of that time. One very intense memory was dinner that my wife and I had with Jan and Mieke Oort on one New Year's Eve. The first snow had fallen, and inside the dining room the white candles lit the blue and white china on the lace tablecloth and were reflected in the silverware. Outside the sky was clear and the stars shone down on the white snow. It was an evening which still gives me pleasure to remember after nearly 20 years.

Another memorable time was the freezing of the canals. Jan Oort is an enthusiastic and first class skater - always the first on the thin ice. The first ice-freeze of the seasons seems to be the only time when astronomy takes second place with him. At the Sterrewacht this occasion warranted a half-holiday with all astronomers on the ice. The unskilled foreigners from hot countries ventured with some reluctance on the slippery black ice, some even using the device employed by small Dutch children, a chair pushed in front of the learner-skater.

During the winter months there were frequent concerts in Leiden, always of a high standard, and if one were not an enthusiastic admirer of J.S. Bach before arrival one would certainly soon have become a Bach lover. Walking home through the quiet streets of Leiden after a concert, with none of the anxieties that might be felt in many cities of the world, was a delightful conclusion to a concert.

It always mystified me that Jan Oort not only takes such an interest in music and the visual arts but has such a wide knowledge of them. His extraordinary dedication to astronomy, stronger than in anyone else that I have ever met, might be expected to leave little time for other things, as in fact it does, but during the relatively short periods between astronomical thinking his enthusiasm for art and for nature seems to be as intense and concentrated as his love of astronomy.

To return to the Benelux Cross - during 1961 and 1962 Oort continued to ponder on what characteristics would be the most suitable for the new telescope. The design favoured by the committee had a number of drawbacks; it could not measure the polarization of incoming radiation, there were no useful line emissions around 400 MHz and the protection from interference at that frequency was doubtful, frequency change would be very difficult, and the relatively long wavelength made necessary antenna dimensions of about 5 km in each direction which would make the finding of a suitable site in the Netherlands very difficult indeed. There were other difficulties too. Some senior Netherlands engineers that Oort would have liked to take a leading part in the project were holding back. More important than even these difficulties was the doubt as to whether the Belgian government would continue to support the project. That government appeared to be interested mainly in the chance of obtaining contracts for Belgian industry, despite the efforts of Belgian scientists to make the government enthusiastic about the scientific advantages to the country.

By 1963 the main design decisions had been made; the Benelux Cross would operate at a wavelength of 21 cm and a modified version of the original parabolic dish design, previously rejected by the committee, would be used. The reduction in wavelength reduced the size of the antenna to less than 1/3 of that of the previous design, polarization measurements would be possible and changes of frequency would be easier. The loss in sensitivity to non-thermal sources in going to the shorter wavelength was to be made up by making the telescope an image forming device, either by producing multiple beams or more radically by extracting all spatial components individually as was done in aperture synthesis. The latter would produce an image of a field about 0.5 degree wide in a very short time, i.e., in a fraction of a second for the Sun, for example, or in a few hours for a very weak source. The new design was a bold one and the telescope would have been extremely powerful had it been built. However, this was not to be, as the Belgian government made it known that it was not willing to proceed any further.

Jan Oort must have been disappointed but not altogether surprised at this setback. Immediately, with the ever-present help of J.H. Bannier of ZWO, he set about finding support for a purely Netherlands telescope. It was obvious that a much more modest design would be needed, and in fact finance was promised for a telescope that would have 10 or 12 parabolic dishes instead of the 100 of the Benelux Cross. With the much-decreased funding of the telescope the main problem was to decide the feature to be sacrificed, high speed of operation or high sensitivity. It was high speed that was dropped - instead of minutes or a few hours to make an observation of a field, several days would now be needed. Some members of the original Benelux Cross Group were still in Leiden and two of them, Hooghoudt and Högbom, produced a new design using elements that were to have been used for the Cross. The new design was a combination of two earth-rotational synthesis instruments, the original

Australian grating telescope and the Cambridge One Mile telescope. Apart from the sacrifice in speed of operation, the new telescope, like the Cambridge one, would be ineffective for observing sources near the celestial equator. However its few elements would make changes of frequency relatively easy and would reduce the complication of equipment needed for observations of the hydrogen line. A further advantage was that the reduction in the number of elements reduced also opposition in the Netherlands to the project, and Oort could now find suitable local people to engage in the work.

The change from the Benelux Cross to the Westerbork Synthesis Radio Telescope was a smooth one, as the new design group was built up around the old one. The principal addition was the building up of first-class data reduction facilities by means of digital computing. The outstandingly fine software designed by Wim Brouw, a student during the early days of the Benelux Cross, was one of the reasons for the immediate success of the Westerbork telescope when it went into action.

To those of us who had been associated with the Benelux project there seemed to be something almost uncanny in the way Oort had steered a course through so many obstacles and near shipwreck and had ended up exactly where he had originally intended to be.

The Westerbork Telescope has been an outstanding success and has made the Netherlands a centre of observational astronomy. The decade since its completion has been one of the most fruitful periods of Jan Oorts long and very fruitful astronomical life.

W.N. Christiansen is retired Professor of Electrical Engineering at Sydney University. He worked several years with the Netherlands Foundation for Radio Astronomy in the Benelux Cross Antenna Project.

W N Christiansen

The Synthesis Radio Telescope at Westerbork.

TEN YEARS OF DISCOVERY WITH OORT'S SYNTHESIS RADIO TELESCOPE

R.J. Allen and R.D. Ekers

1. INTRODUCTION

In 1968 Oort's dicussion (ITR-74) of the priority and astronomical significance of possible programs for the Synthesis Radio Telescope (SRT) covered many topics: surveys of faint radio sources and measurements of their angular sizes, spectra and polarization; determination of the structure and polarization of extended 3C sources; mapping the continuum radiation from "normal" galaxies to determine the dependence on type and the concentration to the centre; investigation of the polarisation of quasi-stellar sources and an attempt to detect the radio-quiet quasi-stellar objects; investigation of individual galaxies in clusters and the cD galaxies in the centre of clusters; studies of radio-weak Seyfert galaxies and galactic X-ray sources with the great sensitivity of the SRT; and the determination of the general distribution of HI and its relationship to optically visible structures in galaxies. In the following collection of the most significant contributions the SRT has made to astronomy we can see the development of many of these ideas.

We have made no attempt to review all the fields of astronomy touched by the SRT. We have also sacrificed completeness in order to assemble here what seem to us to be the most outstanding contributions. Where possible this material is presented by illustrations, since an important impact of the SRT has been its power to map the radio sky with optical detail. We have tried to select the illustrations as close as possible to the first results, in order to recover some of the excitement of these discoveries; in some cases there have been more detailed reductions or follow-up observations which we have not included.

2. GALACTIC NUCLEI, RADIO GALAXIES AND COSMOLOGY

An underlying theme in many of the SRT projects has been the role of the activity in the nucleus of galaxies. *In de laatste twee decennia is het geleidelijk duidelijker geworden dat in de kernen van vele - wellicht zelfs van de meeste - stelsels een soort aktiviteit plaatsvindt*

H. van Woerden, W. N. Brouw, and H. C. van de Hulst (eds.), Oort and the Universe, 79–110.

die af en toe tot bijzonder hevige eruptieve verschijnselen leidt. Welke enorme omvang deze 'explosiviteit' kan aannemen, blijkt in de zgn. quasars, waarin de kern de hele rest van het sterrenstelsel ver overstraalt. (Oort 1970). The SRT observations have ranged from the nucleus of our Galaxy and other normal galaxies to Seyfert galaxies, radio galaxies and quasars.

Despite the unfavourable declination of the nucleus of our Galaxy the SRT, in conjunction with the Owens Valley Telescope, produced the first map of the entire Sgr A source (WORST map, Fig. 1). In this map the nature of the two components Sgr A East and West was clarified. The SRT did not have enough resolution to separate clearly the ultra-compact source embedded in Sgr A West; however, it was used to discover similar flat-spectrum compact sources in two other spiral galaxies, M81 and M104 (the Sombrero Nebula) (Ekers 1975), and to show that these two radio nuclei were varying in intensity (de Bruyn et al. 1976). For the majority of normal galaxies the SRT has found a rather complex distribution of nonthermal emission from the nucleus. Relatively few clear relations to other properties have been found, but one of the early discoveries which is now well confirmed was the correlation between infrared emission and radio flux density shown in Fig. 2. Another result coming from a recent systematic survey of bright galaxies with the SRT has shown that the nuclear emission, but not the disk emission, is enhanced when there are other galaxies in the neighbourhood (Fig. 3).

Moving up to higher-energy phenomena, the SRT has found some quite dramatic effects of the nucleus. The first SRT observation of the head-tail shaped radio sources, which had been discovered at Cambridge a few years earlier, gave striking new clues to the nature of these sources and of radio galaxies in general. As the Leiden photographer Mr. Kleibrink developed Fig. 4, he saw a reminder of the trail of a World War II bomber, and for every radio astronomer the message of the trail of radio-emitting debris has been just as clear. Even more dramatic perspective was added to this picture by one of the first 6-cm SRT observations of another head-tail source, NGC1265 (Fig. 5). This lays before us the entire picture, from the ejection of blobs from the nucleus of the elliptical galaxy, to the formation of the diffuse regions in the tail. Observations of the normal double-lobed radio galaxies with increasing resolution (Fig. 6) showed that a similar jet-like stream of radio plasma was also embedded in the large-scale structure, essentially confirming the main idea of the beam-type models of radio galaxies. How these models can cope with the requirements such as those imposed by the precession of this beam, as suggested by sources such as that in Fig. 7, remains to be seen.

While the new 6-cm SRT was unravelling some of the finest-scale structures seen in radio galaxies, the piggy-backed 50-cm system was used to map the radio galaxy DA240 and thereby discover the largest object in the Universe (Fig. 8). Even this record was soon bettered by an even larger radio galaxy, 3C236 (Fig. 9). A VLBI observation using the SRT and other European telescopes shows the alignment between the nuclear radio source and the outer lobes.

Other large-scale objects for which the SRT has been uniquely suited are the great clusters of galaxies. Fig. 10 is the first map which showed the smooth centrally concentrated structure of Coma C, a well-known but never-mapped extended source in the centre of the Coma cluster. Perhaps related is the chaotic mixture of tails and radio lobes found in the centre of Abell 2256 (Fig. 11).

Both of these figures also show the large number of field sources typical for any observation with the SRT. A few of these belong to the clusters, but most are background objects at much greater distances and as such provide a tool for investigating the Universe on the largest possible scale. Counts and the measurement of the angular size of these weak radio sources have extended both the source counts and the angular size-flux density relations a factor of ten fainter than previous surveys (Figs. 12 and 13). Both these relations deviate strongly from any non-evolving cosmological model, and their interpretation requires an understanding of how the average density and perhaps sizes of radio sources have changed with cosmic epoch. The first direct measurement of this evolution has come from the identification of faint SRT field sources on plates from the Palomar 48-inch and Kitt Peak 4-meter telescopes. Fig. 14 shows a density increase by a factor of 10 to 100 over the local value.

Gravitational theory itself was the subject of successful SRT experiments which measured the deflection of radio sources in the Sun's gravity field and confirmed Einstein's theory to an accuracy of 4%. The unique observing technique is illustrated in Fig. 15a. This was expanded to include two frequencies and gave the result in Fig. 15b. Subsequent experiments of the same type at NRAO have now increased the accuracy of this type of measurement to 1%.

3. STRUCTURE AND DYNAMICS OF GALAXIES

While planning the first observing programs for the SRT, Oort (1968) also emphasized the study of HI and continuum spiral structure and of its relationship to the optical tracers in spiral galaxies. The spectacular results on the 21-cm continuum in M51 followed scarcely two years later (Fig. 16), serving to confirm once more Oort's scientific perspicacity. This figure was the form in which the M51 observation was first seen during morning coffee at the Sterrewacht te Leiden - surely a memorable morning. Subsequent detailed comparisons with the dust, HI, Hα, and radio-continuum polarization have led to fundamental results on the degree of gas compression and the ordering of the magnetic field, and on the time scales for star formation in spiral arms.

The next (and only other) galaxy found to have a strong spiral morphology in the continuum was NGC4258 (Fig. 17) and, although there did seem to be some enhancement of the radio emission along the normal spiral arms, another more powerful mechanism was clearly at work. The model which was proposed reflected one of Oort's basic convictions that .....*the ultimate source of nonthermal radiation of galaxies is likely to lie in their nuclei* (Oort, 1967).

The provisional conversion of the SRT into a synthesis spectrometer in 1971 opened new possibilities for the study of the dynamics of the nearby galaxies. The first results on M101 (Fig. 18) provided an unequivocal example of the close correspondence between the optical spiral structure and the distribution of HI. The SRT spectrometer has yielded an immense amount of information; as an example, the results on M101 alone have led to ten publications and are still being further exploited today. Perhaps the crowning achievement of the SRT in the study of spiral structure is the work on M81, which demonstrated a detailed dynamical correspondence of the observations with the major features of the density-wave theory (Fig. 19).

The SRT has also provided fundamental new insights into the general distribution of matter in R and z in normal galaxies. The degree of detail in the results can be illustrated with the spectral-index distribution in NGC6946 (Fig. 20); it remains a curious unexplained result that the radial dependences of the nonthermal radio and the optical surface brightness correspond so well. At greater distances from the nucleus, the observed rapid changes in spectral index can be interpreted as changes in the electron energies and magnetic-field strengths near the "edges" of the galaxy. The general morphology of HI and continuum in NGC891 (Fig. 22) shows the strong concentration of the HI to the central plane of the galaxy, whereas the continuum is much thicker. The observations of NGC4631 (Fig. 21) gave the first clear evidence for a radio-continuum halo in a normal galaxy, and the HI distribution in NGC5907 (Fig. 23) showed a warp out of the central plane with a degree of clarity which escaped many years of research in our own Galaxy.

The number and variety of galaxies observed in the HI-line with the SRT slowly grew through the decade, and near the end of the 70's it became possible to study the relationship of dynamical parameters such as the distribution of mass and mass-to-luminosity ratio for galaxies of different types. A collection of HI radial-velocity fields, mostly obtained with the SRT, is shown in Fig. 24. These results gave a clear indication for an exponential increase of the M/L ratio with distance from the centre of a galaxy (Fig. 25). However, the identification of the important physical parameters underlying the optical morphological types remains elusive. The enigmatic HI in elliptical galaxies has also come into the reach of the SRT. The gas in NGC4278 has a size similar to the optical but the motions are not consistent with the optical rotation (Fig. 26).

The great sensitivity of the SRT for high-resolution studies of faint surface-brightness regions yielded surprising new results on the effects of interactions of normal galaxies with their neighbours. First suspected of being embedded in an "envelope" of HI, M81 turned out to be surrounded with filaments (Fig. 27) which had a clear relation to the inner spiral structure and to a recent close passage of NGC3077. The enigma of Stephan's Quintet may be closer to an explanation through the discovery with the SRT of a galaxy-sized mass of HI (Fig. 28) separated from all the bright optical objects in the field; the evidence is in favour of a recent violent interaction of several high-redshift members of the group. Finally the SRT results on NGC2685 (Fig. 29) suggest that large amounts of HI gas are being accreted from the surroundings, producing dynamical chaos in the galaxy.

4. SOLAR SYSTEM, STELLAR AND INTERSTELLAR ASTRONOMY

The speed of the SRT made it possible to obtain observations of moving or transient phenomena. These have included solar events and the radio emission from Jupiter which has been mapped in all Stokes parameters as a function of Jupiter's longitude (Fig. 30).

Observations of objects within our Galaxy also received attention in Oort's early proposals for SRT observing programs, where he predicted that *Because of its great sensitivity for small sources, the SRT is most suited to search for radio-frequency radiation of special objects like X-ray sources.* The first discovery in this area with the SRT was in its own way almost as spectacular as the map of M51. One of the discrete sources in a field in the Cygnus region near the position of the X-ray source Cygnus X-3 showed grating responses in places where none were expected (Fig. 31), indicating a strongly variable radio emission from the X-ray object.

The high resolution and sensitivity of the SRT has also revealed features in small, compact galactic HII-regions thought to be closely associated with active star formation. The example of NGC7538 (Fig. 32) is typical of such young, evolving HII-regions of high density with associated OH and H_2O maser sources and infrared objects. In this same HII-region the first H_2CO-maser has recently been discovered with the SRT (Forster et al. 1980). SRT synthesis of recombination lines at 6 cm (Fig. 33) has also produced detailed information on the kinematics and variations of electron temperature within HII-regions.

One of the early SRT observations of galactic filamentary nebulae refuted the long-standing hypothesis that obscuration by dust played a major role in defining their optical appearance; the remarkable correspondence between the 21-cm continuum and the optical emission lines in the filamentary nebula NGC6888 (Fig. 34) demonstrated that the thermal emission itself is filamentary.

A subject which has long held Oort's interest is the origin of the high-velocity clouds originally discovered in galactic HI surveys at Dwingeloo. Here the good angular and frequency resolution of the SRT with the new digital line receiver has shown a fine filamentary structure (Fig. 35), which will surely provide new insight into these enigmatic clouds of neutral gas.

As a final example of SRT discoveries in galactic astronomy, Fig. 36 shows the recent mapping of proper motion in the expanding shell of Tycho's supernova of 1572.

EPILOGUE

Jan Oort's wisdom and foresight have been a source of inspiration to all of us who have had the privilege of his guidance and interest during the many programs of observation with the SRT and the astrophysical analysis of the results. His original vision and the dedication of the engineers and support staff of the Netherlands Foundation for Radio Astronomy have provided us, the astronomers, with one of the world's most powerful and accurate radio-astronomical instruments during ten exciting years of discovery.

References

Allen, R.J., Goss, W.M. and Van Woerden, H. 1973, Astron. Astrophys. 29, pp. 447.

Allen, R.J. and Sullivan, W.T. III 1980, Astron. Astrophys. 84, pp. 181.

Bosma, A. 1978, "The Distribution and Kinematics of Neutral Hydrogen in Spiral Galaxies of various Morphological Types", Ph. D. Thesis, University of Groningen.

Braes, L.L.E. and Miley, G.K. 1972, Nature 237, pp. 506.

Bridle, A.H. and Fomalont, E.B. 1976, Astron. Astrophys. 52, pp. 107.

De Bruyn, A.G., Crane, P.C., Price, R.M. and Carlson, J.B. 1976, Astron. Astrophys. 46, pp. 243.

De Pater, I. 1980, submitted to Astron. Astrophys.

De Ruiter, H.R. 1978, "Faint Extra-Galactic Radio Sources and their Optical Identifications", Ph. D. Thesis, University of Leiden.

Ekers, R.D. and Sancisi, R. 1977, Astron. Astrophys. 54, pp. 973.

Ekers, R.D. 1975, "Structure and Evolution of Galaxies" (ed. G. Setti), Reidel, Dordrecht, pp. 217.

Ekers, R.D., Goss, W.M., Schwarz, U.J., Downes, D. and Rogstad, D.H. 1975, Astron. Astrophys. 43, pp. 159.

Ekers, R.D. and Miley, G.K. 1977, "Radio Astronomy and Cosmology" (ed. D.L. Jauncey), IAU Symp. 74, pp. 109.

Ekers, R.D., Fanti, R., Lari, C. and Parma, P. 1978, Nature 276, pp. 588.

Forster, J.R., Goss, W.M., Wilson, T.L., Downes, D. and Dickel, H.R. 1980, Astron. Astrophys. 84, L1.

Hummel, E. 1980, "The Radio Continuum Structure of Bright Galaxies at 1.4 GHz", Ph. D. Thesis, University of Groningen.

Israel, F.P., Habing, H.J. and De Jong, T. 1973, Astron. Astrophys. 27, pp. 143.

Jaffe, W.J., Perola, G.C. and Valentijn, E.A. 1976, Astron. Astrophys. 49, pp. 179.

Katgert, P. 1975, Astron. Astrophys. 38, pp. 87.

Maslowski, J. 1977, "Radio Astronomy and Cosmology" (ed. D.L. Jauncey), IAU Symp. 74, pp. 49.

Miley, G.K., Perola, G.C., Van der Kruit, P.C. and Van der Laan, H. 1972, Nature 237, pp. 269.

Oort, J.H. 1967, Internal Technical Report 49, Netherlands Foundation for Radio Astronomy.

Oort, J.H. 1968, Internal Technical Report 74, Netherlands Foundation for Radio Astronomy.

Oort, J.H. 1971, Yearly Report of the Netherlands Foundation for Radio Astronomy for 1970, pp. 28.

Raimond, E., Faber, S.M., Gallagher, J.S. and Knapp, G.R. 1980, Astrophys. J., in preparation.

Rots, A.H. 1974, "Distribution and Kinematics of Neutral Hydrogen in the Spiral Galaxy M81", Ph. D. Thesis, University of Groningen.

Sancisi, R. 1976, Astron. Astrophys. 53, pp. 159.

Sancisi, R. and Allen, R.J. 1979, Astron. Astrophys. 74, pp. 73.

Shane, W.W. 1980, Astron. Astrophys. 82, pp. 314.

Van Breugel, W.J.M. and Miley, G.K. 1977, Nature 265, pp. 315.

Van Gorkom, J., Goss, W.M., Shaver, P.A. and Harten, R.H. 1980, Yearly Report of the Netherlands Foundation for Radio Astronomy 1979, pp.90.
Van der Hulst, J.M. 1977, "The Distribution and Motions of Neutral Hydrogen in the Interacting Galaxy Pairs NGC4038/39 and NGC3031/77", Ph. D. Thesis, University of Groningen.
Van der Kruit, P.C. 1971, Astron. Astrophys. 15, pp. 110.
Van der Kruit, P.C., Oort, J.H. and Mathewson, D.S. 1972, Astron. Astrophys. 21, pp. 169.
Van der Kruit, P.C. 1973, Astron. Astrophys. 29, pp. 263.
Van der Kruit, P.C., Allen, R.J. and Rots, A.H. 1977, Astron. Astrophys. 55, pp. 421.
Visser, H.C.D. 1978, "The Dynamics of the Spiral Galaxy M81", Ph. D. Thesis, University of Groningen.
Weiler, K.W., Ekers, R.D., Raimond, E. and Wellington, K.J. 1974, Astron. Astrophys. 30, pp. 241.
Weiler, K.W., Ekers, R.D., Raimond, E. and Wellington, K.J. 1975, Physical Review Letters 35, pp. 134.
Wellington, K.J., Miley, G.K. and Van der Laan, H. 1973, Nature 299, pp. 502.
Wendker, H.J., Smith, L.F., Israel, F.P., Habing, H.J. and Dickel, H.R. 1975, Astron. Astrophys. 42, pp. 173.
Willis, A.G., Strom, R.G. and Wilson, A.S. 1974, Nature 250, pp. 625.

R.J. Allen studied at the University of Saskatchewan and at Massachusetts Institute of Technology; R.D. Ekers studied at the University of Adelaide and at Australian National University. Allen and Ekers have been at the Kapteyn Institute in Groningen since 1969 and 1971, respectively; both are Professor of Radio Astronomy.

R J Allen. R. D. Ekers

Thanks are due to the publishers of Nature, Astronomy and Astrophysics, and Physical Review for permission to reproduce Figures 4-8, 10, and 16.

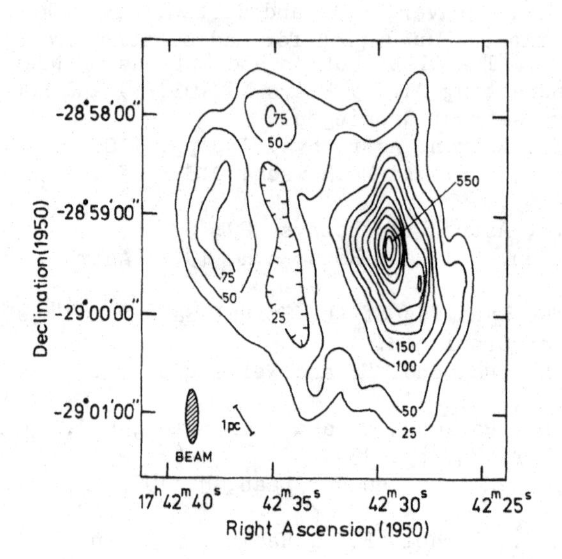

1 Sgr A mapped at 6 cm; combination of data from Westerbork and Owens Valley Radio Synthesis Telescopes (WORST). From Ekers, Goss, Schwarz Downes and Rogstad (1975).

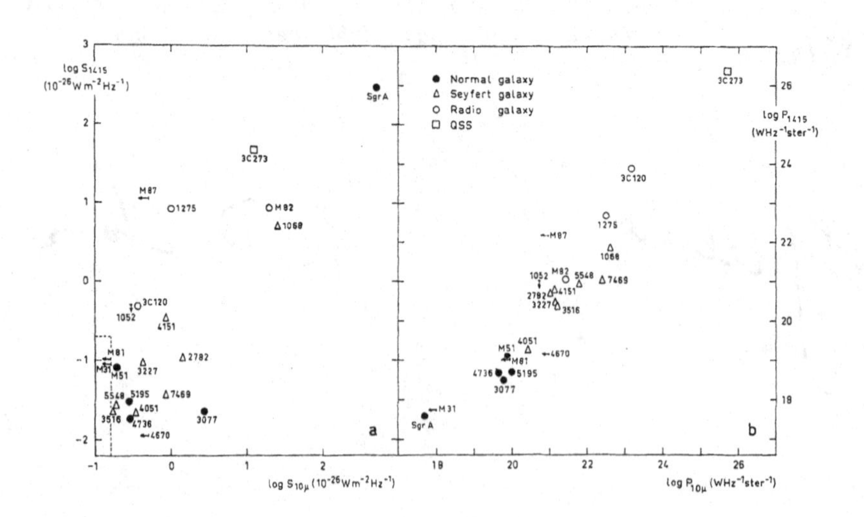

2 Plots showing a correlation between 10-μ and 21-cm emission from nuclei of galaxies for (a) flux density and (b) power. From Van der Kruit (1971).

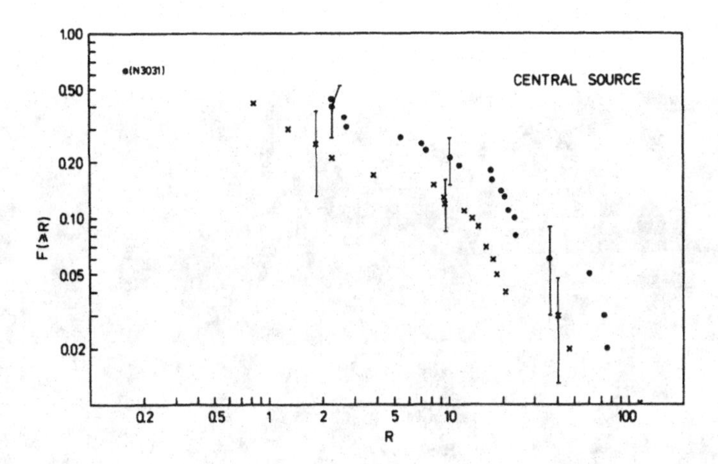

3 The radio/optical luminosity function for the central sources of spiral galaxies showing the difference between isolated galaxies (x) and those in multiple systems(.). From Hummel (1980).

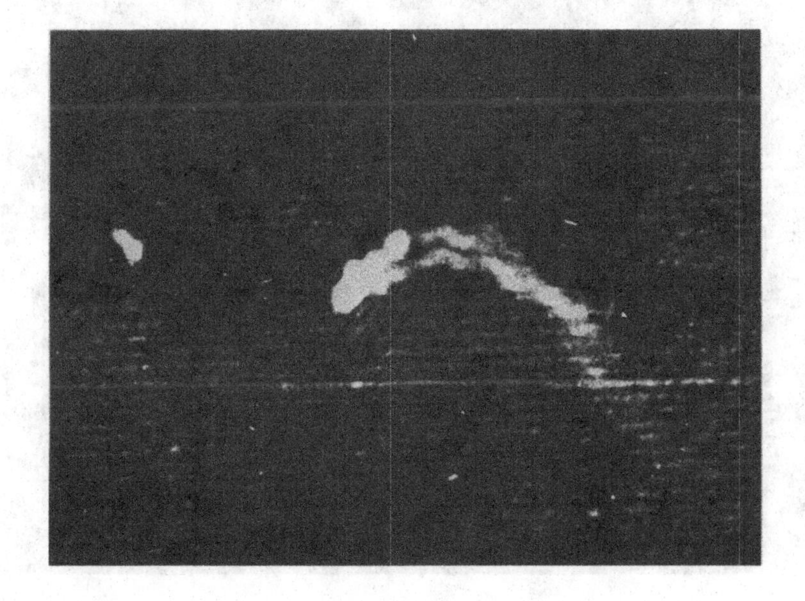

4 The radio galaxy 3C129 mapped at 21 cm. Described in Miley, Perola, Van der Kruit and Van der Laan (1972).

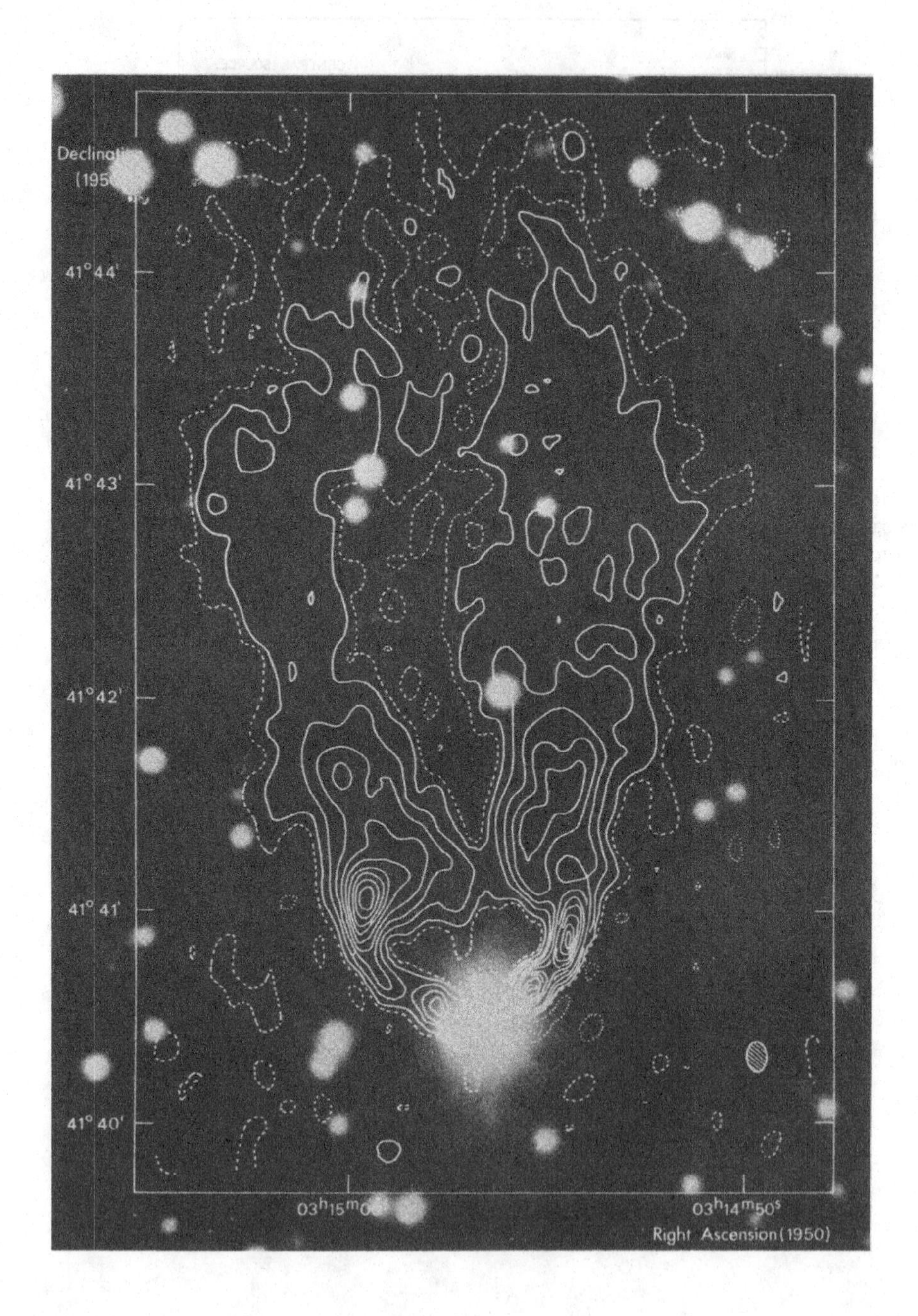

5 . The 6-cm emission from NGC1265 superimposed on an optical photograph. From Wellington, Miley and Van der Laan (1973).

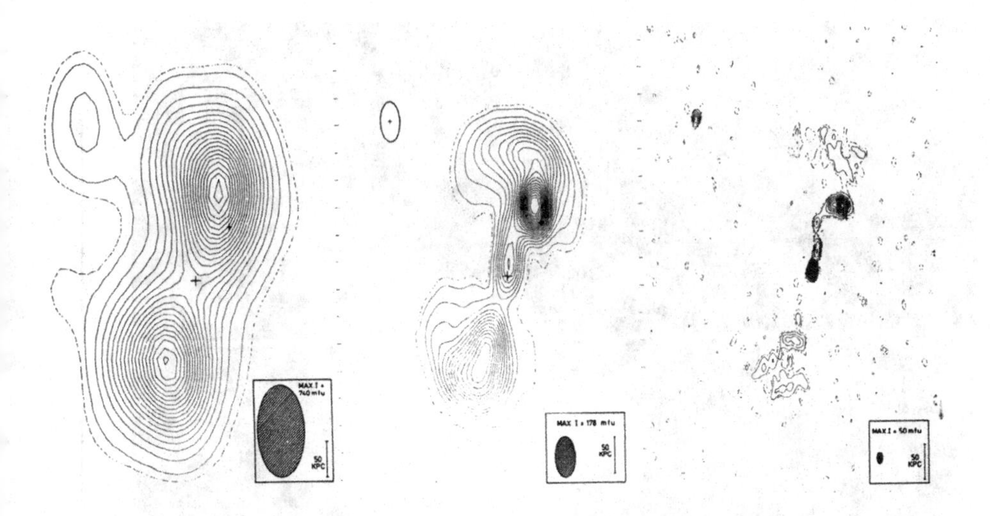

6 The radio galaxy B 0844+31 at (a) 50 cm, (b) 21 cm and (c) 6 cm. From Van Breugel and Miley (1977).

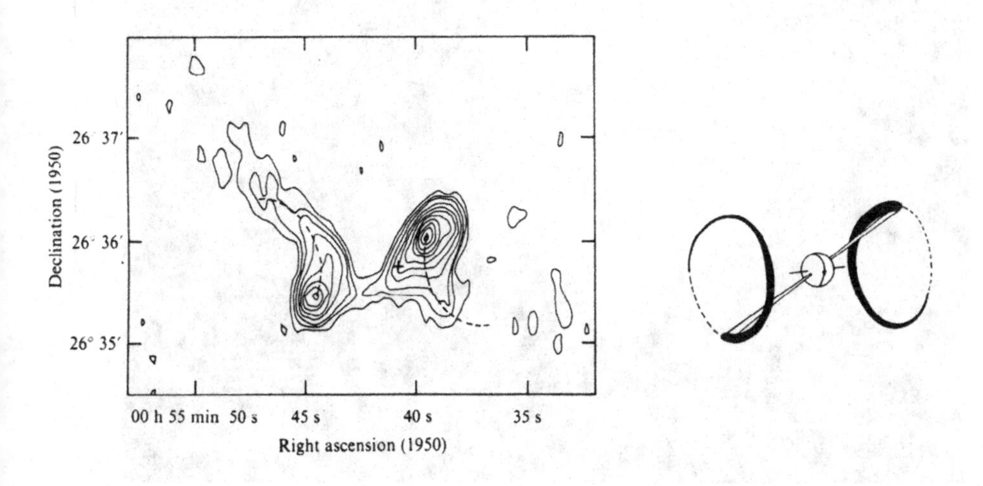

7 The radio galaxy B2 0055+26 (NGC326) (left) at 6 cm and (right) the proposed model. From Ekers, Fanti, Lari and Parma (1978).

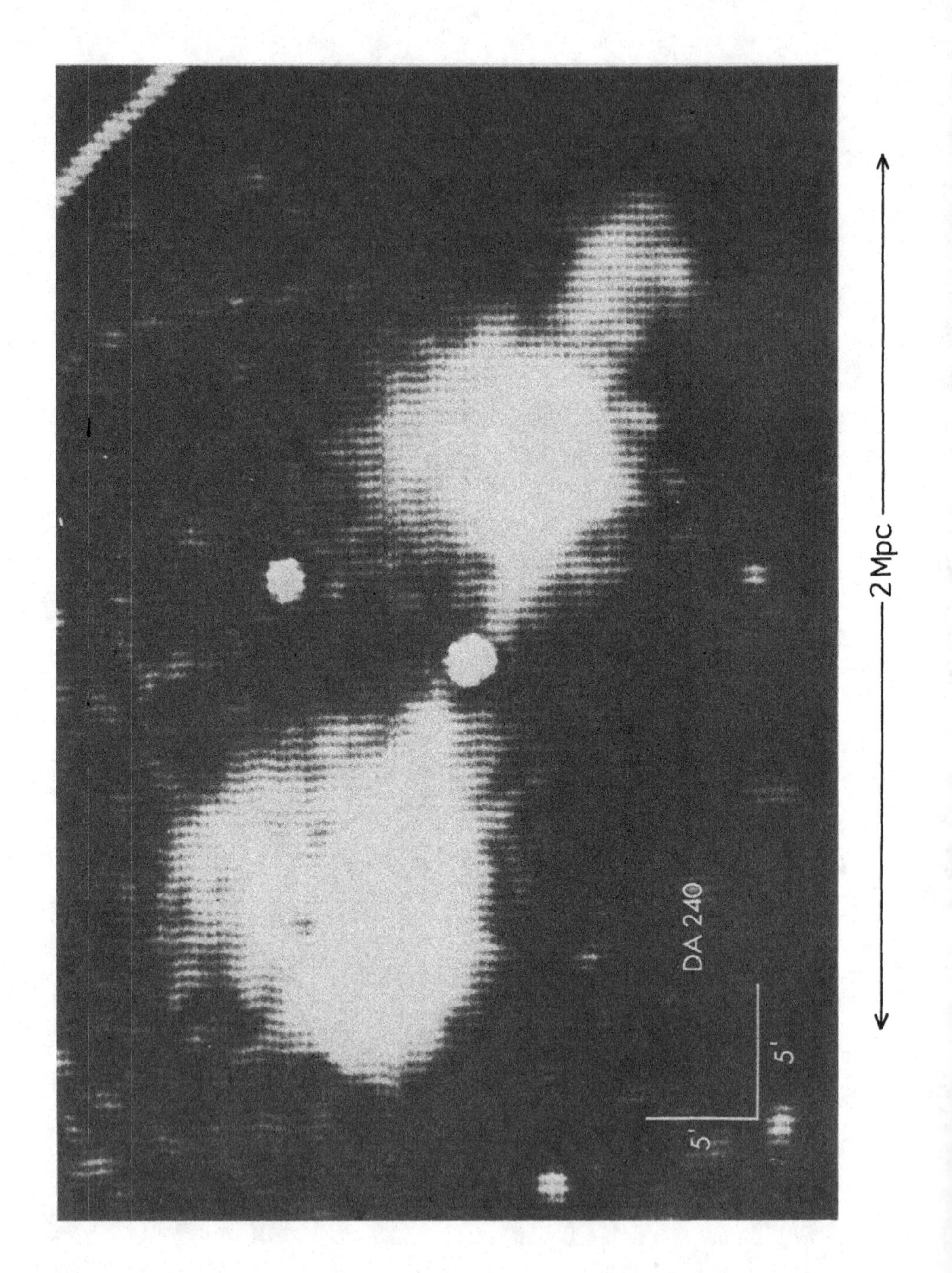

8 The giant radio galaxy DA240 at 50 cm. From Willis, Strom and Wilson (1974).

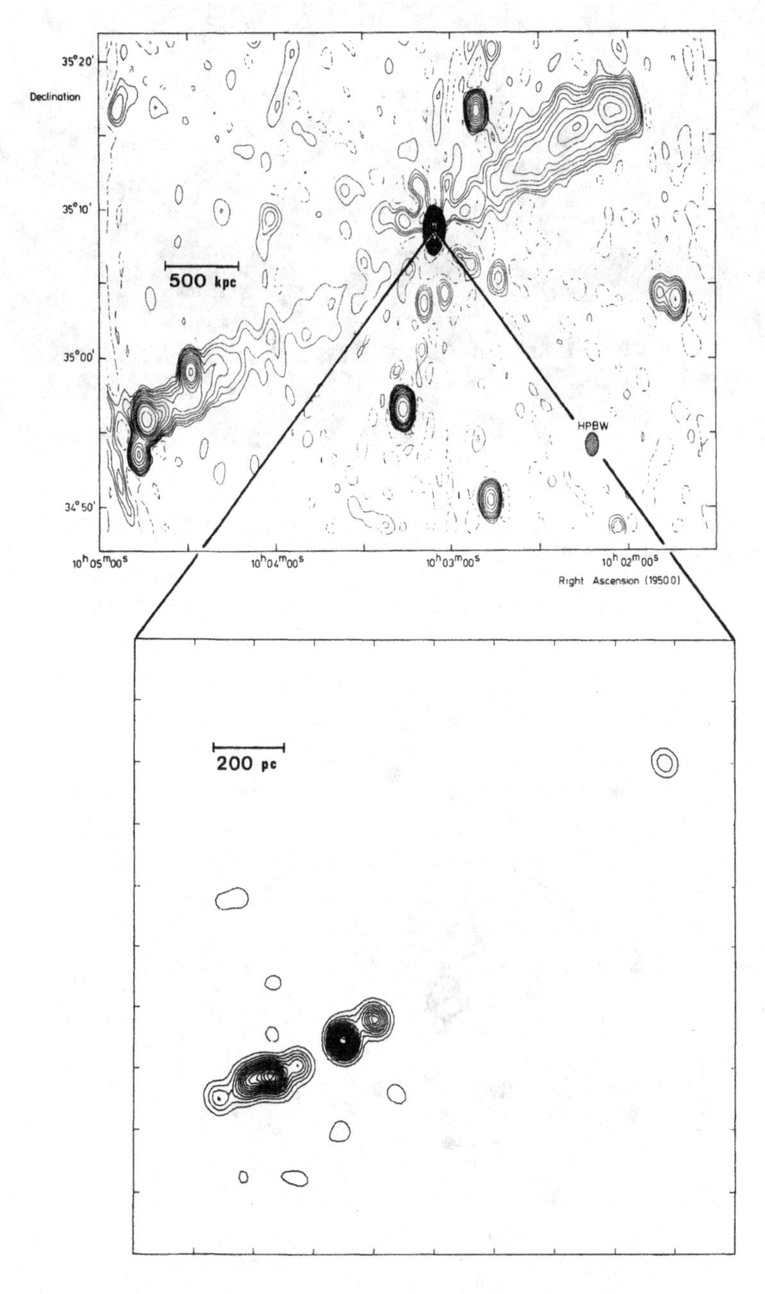

9 (top) The giant radio galaxy 3C236 at 50 cm. From Willis, Strom and Wilson (1974).
(bottom) VLBI observation of the central component. From Schilizzi (private communication).

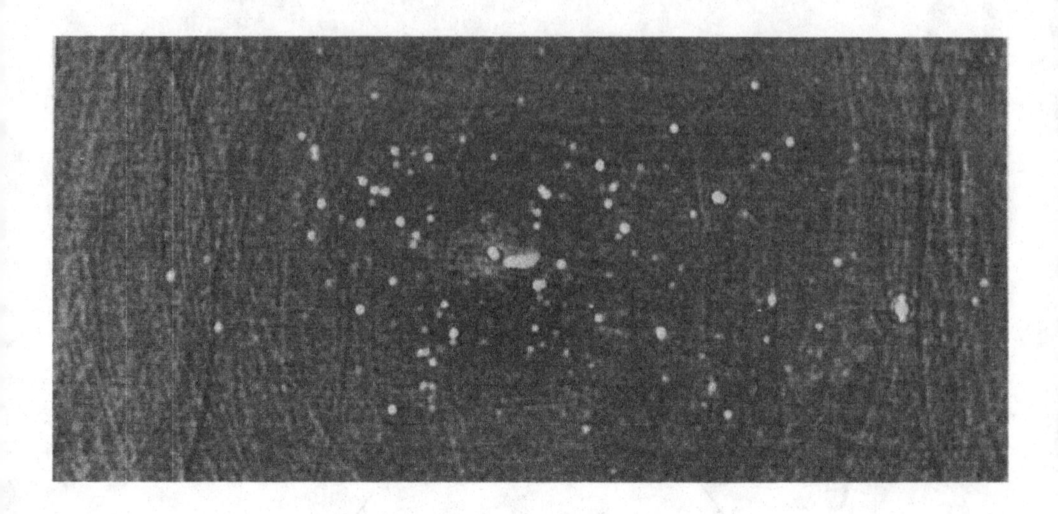

10 50 cm emission from the great galaxy cluster in Coma. Described in Jaffe, Perola and Valentijn (1976).

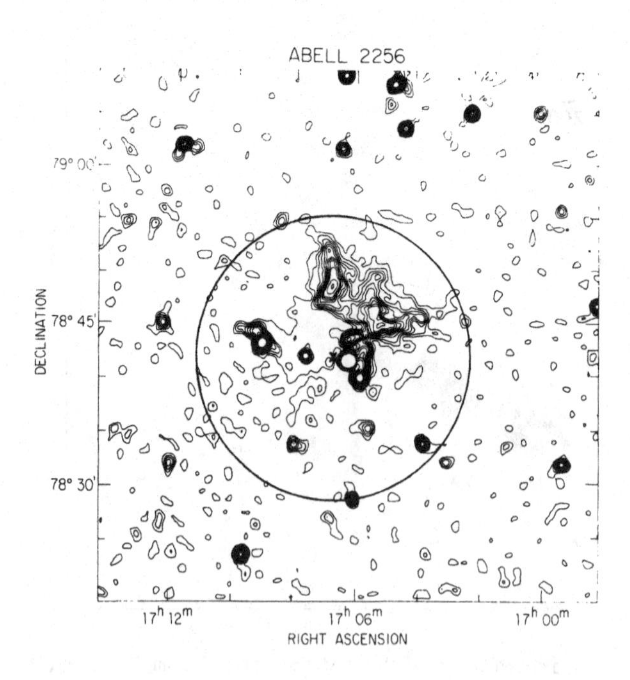

11 The galaxy cluster Abell 2256 at 50 cm. The circle indicates the size of the cluster. From Bridle and Fomalont (1976).

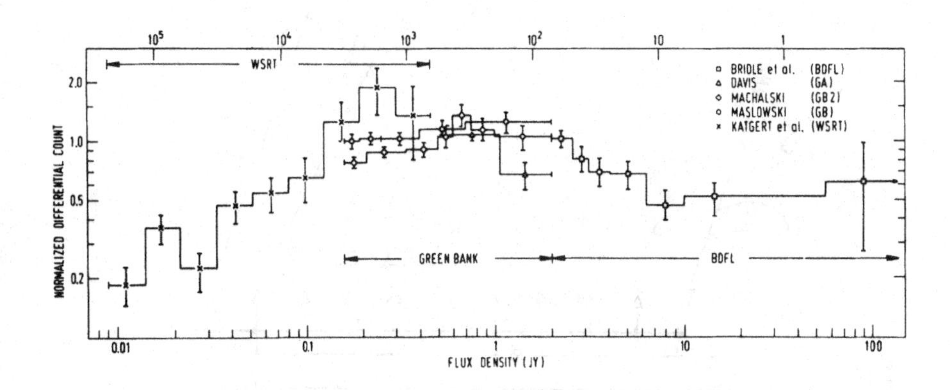

12 The density of radio sources (normalised by the Euclidean value) as a function of flux density at 21 cm. From a compilation of data by Maslowski (1977). This SRT data is from Katgert (1975).

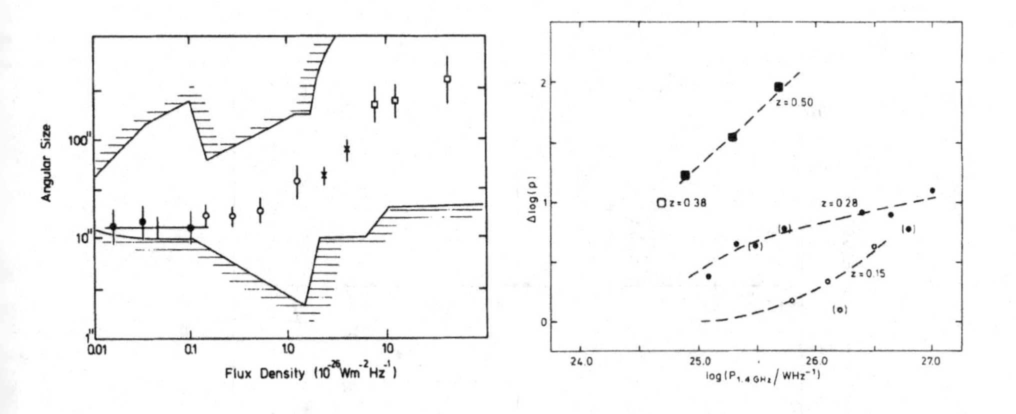

13 Median angular size of radio sources as a function of flux density using a combination of data; ● WSRT, O Ooty, X BDFL, □ All Sky. From Ekers and Miley (1977) (left panel).

14 The space density of radio sources compared with the local density as a function of power at 21 cm for various epochs in the past. From De Ruiter (1978) (right panel).

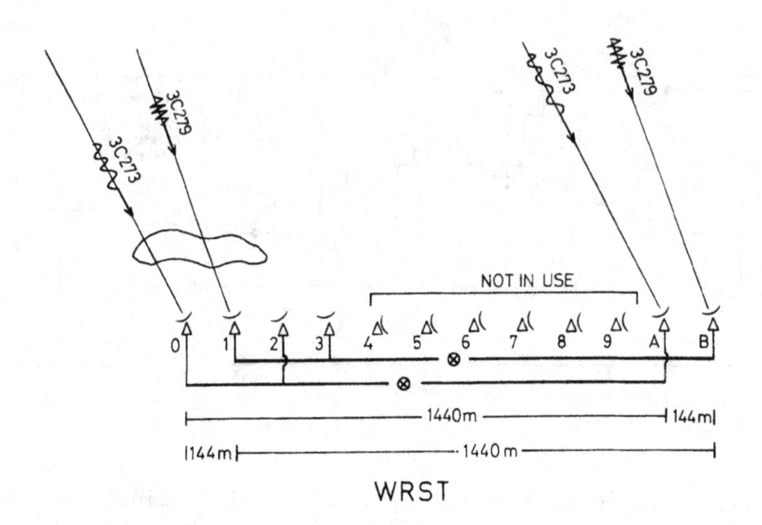

15 (a) Schematic illustration of the technique used to observe the solar gravitation (above), and (b) the results obtained as the quasar 3C279 passed the sun (below). From Weiler, Ekers, Raimond and Wellington (1974 and 1975).

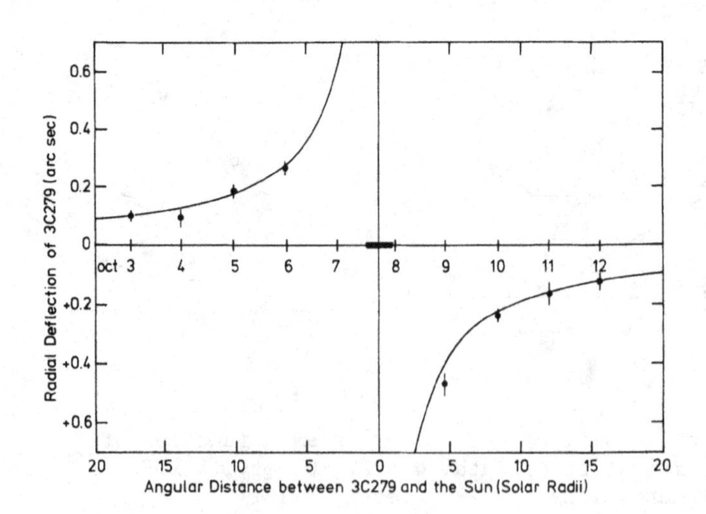

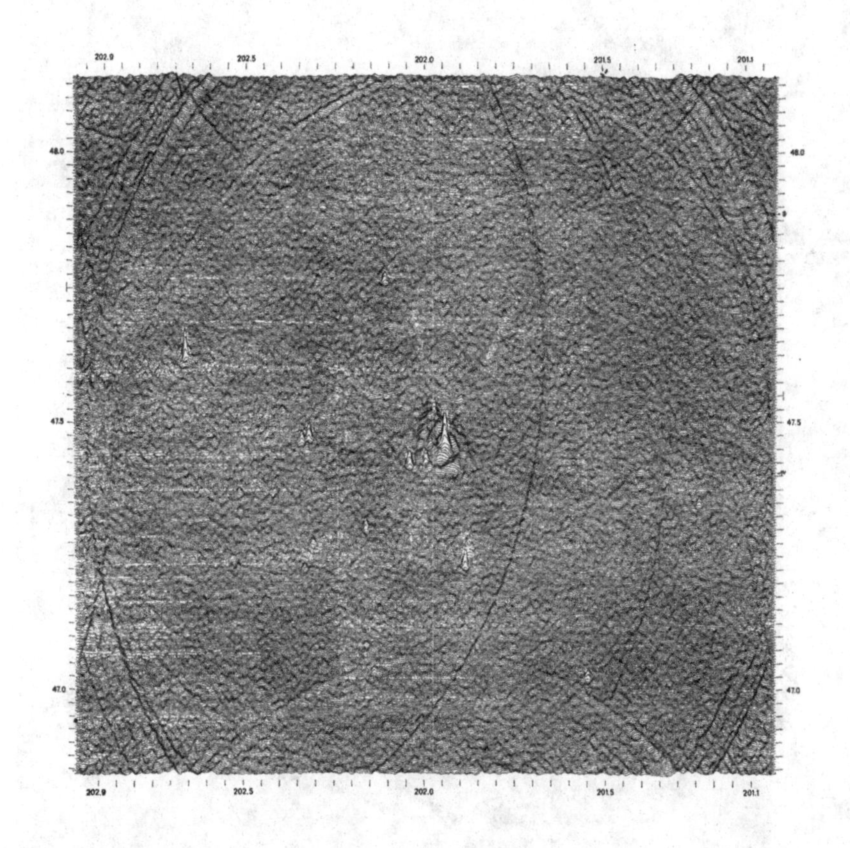

16 4x12 hr continuum synthesis of the M51 field (1°.3 x 1°.3) with the SRT at 1415 MHz showing the nuclear source and the spiral arms. From Oort (1971).

17 Radio contour map of NGC4258 at 1415 MHz superposed on an optical photograph of the galaxy. From Van der Kruit, Oort and Mathewson (1972).

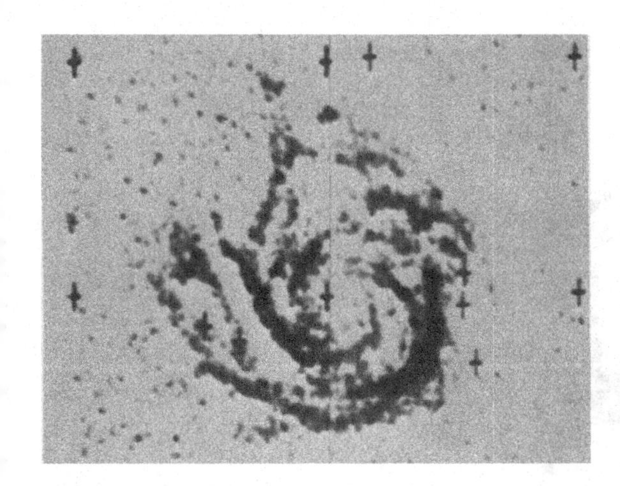

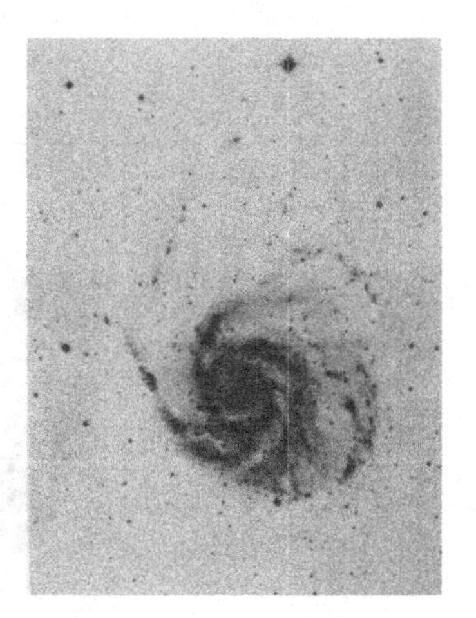

18 Distribution of the HI in M101 (left panel) compared to the optical spiral structure (right panel). From Allen, Goss and Van Woerden (1973).

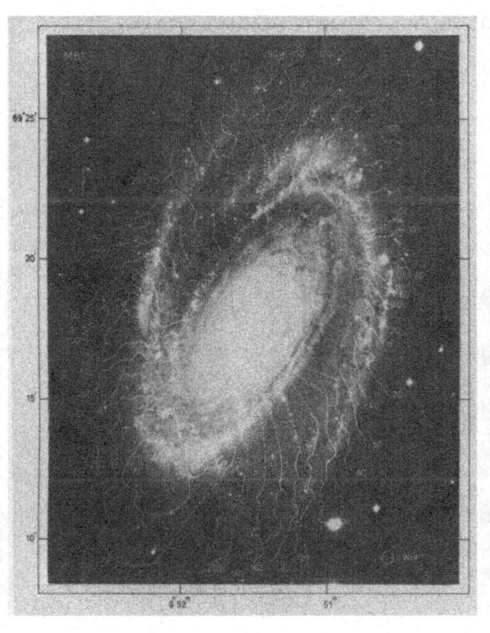

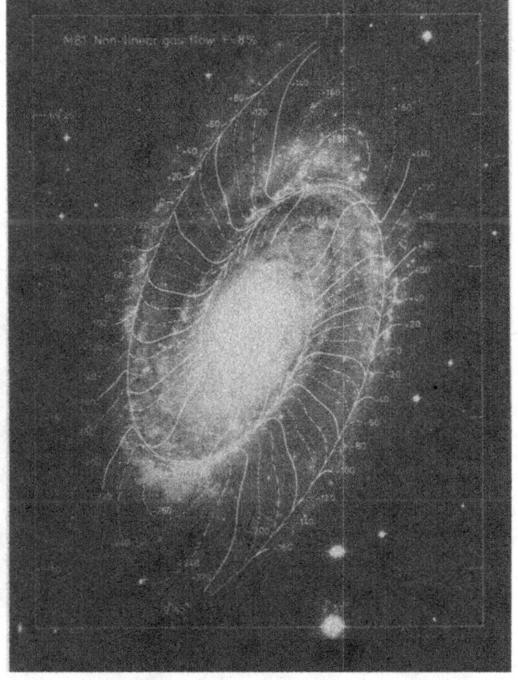

19 The HI radial velocity field of M81 superposed on an optical photograph (left panel) compared to the prediction of the density-wave theory (right panel). Adapted from the earlier work of Rots (1974) and Shane, and refined by Visser (1978).

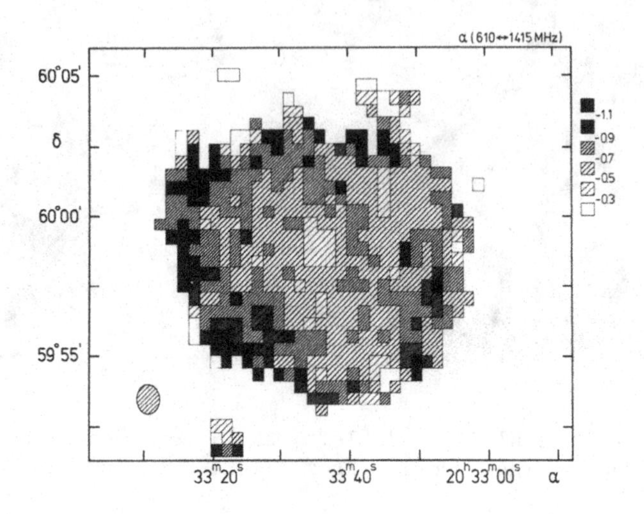

20 Distribution of spectral index in NGC6946. From Van der Kruit, Allen and Rots (1977).

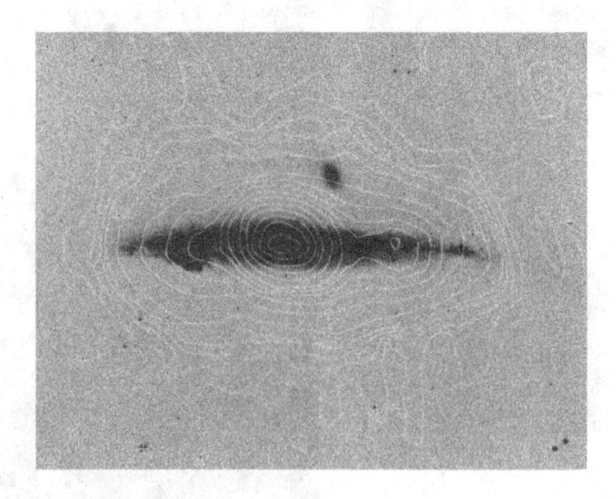

21 49-cm continuum map of NGC4631 showing the z-extent into a halo. From Ekers and Sancisi (1977).

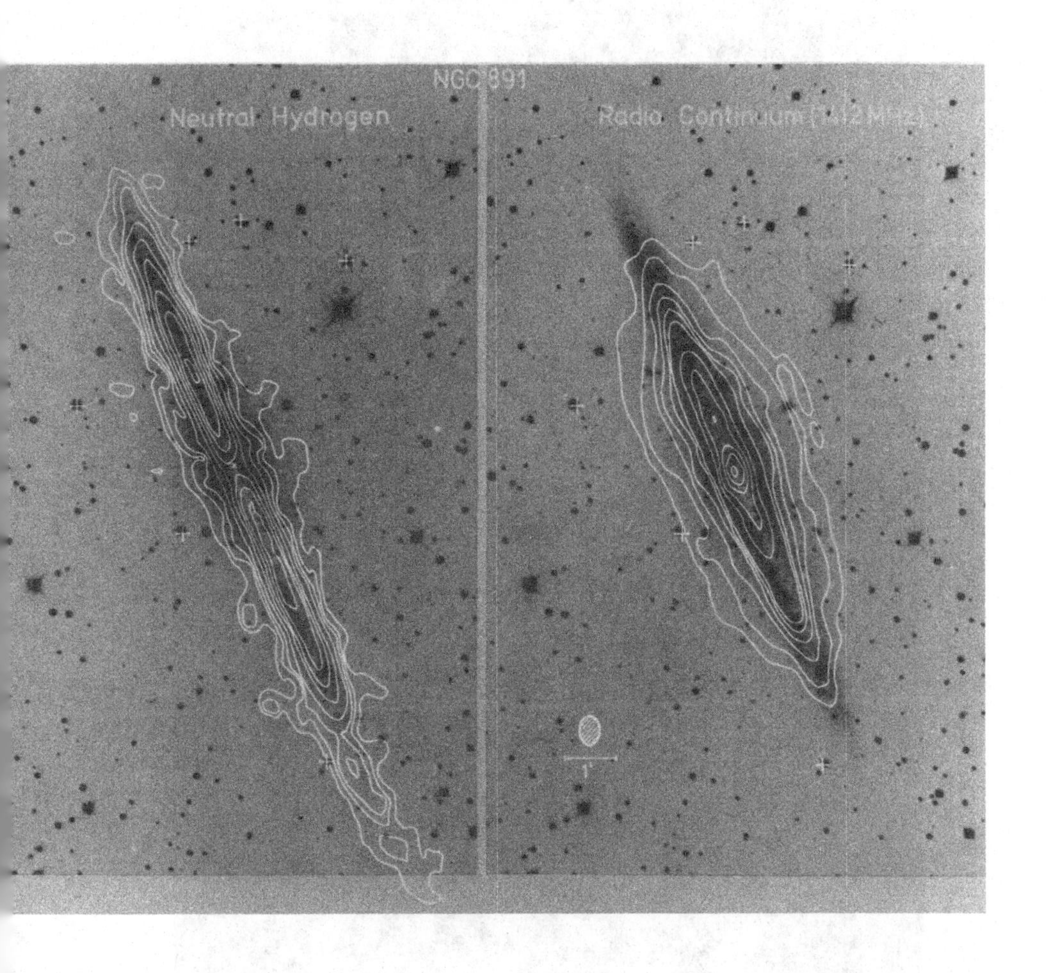

22 Distribution of HI (left panel) and 21-cm radio continuum (right panel) in NGC891. From Sancisi and Allen (1979).

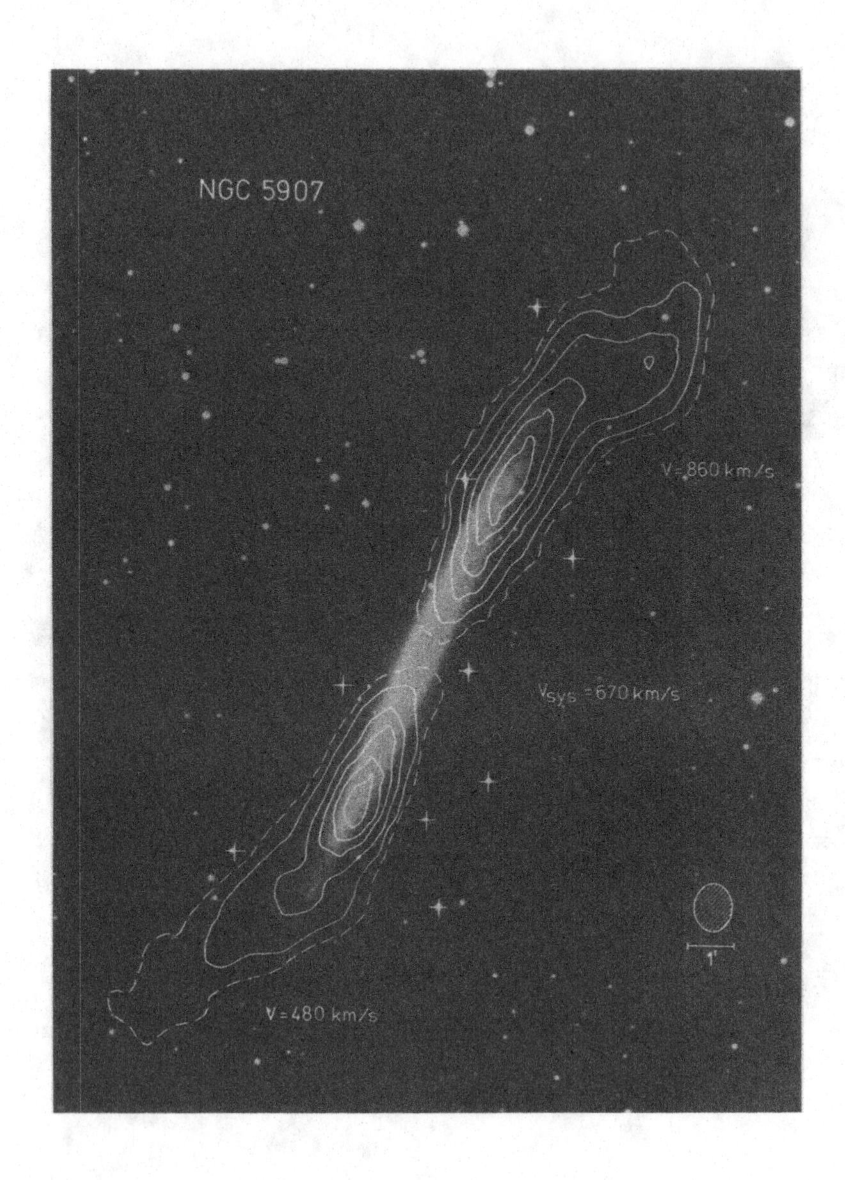

23 HI channel maps at two extreme radial velocities in NGC5907. From Sancisi (1976).

24 Radial velocity fields of 22 spiral galaxies. From Bosma (1978).

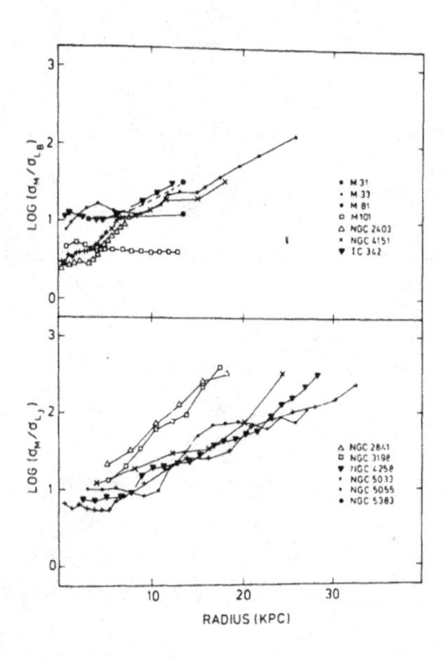

25 Surface density ratios as a function of radius from the centre for a number of nearby galaxies. From Bosma (1978).

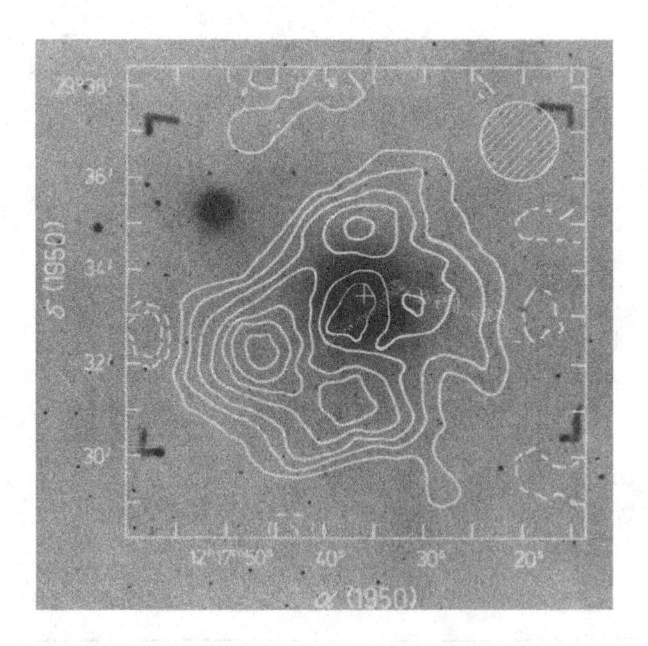

26 HI distribution in NGC4278. The kinematical major axis of the HI has a position angle 45° different from the major axis of the light distribution. From Raimond, Faber, Gallagher and Knapp (1980).

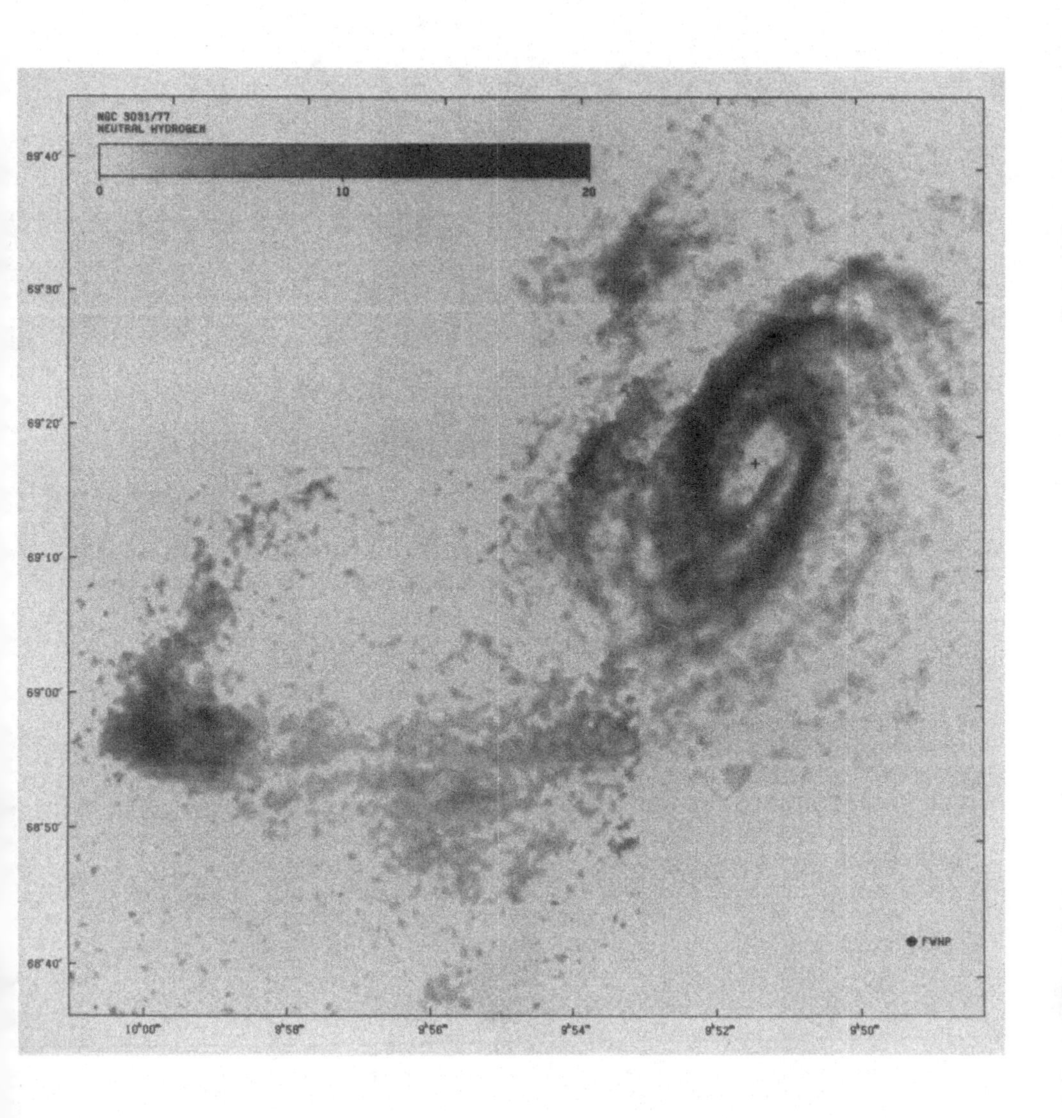

27 HI distribution in the region of M81 and NGC3077. From Van der Hulst (1977).

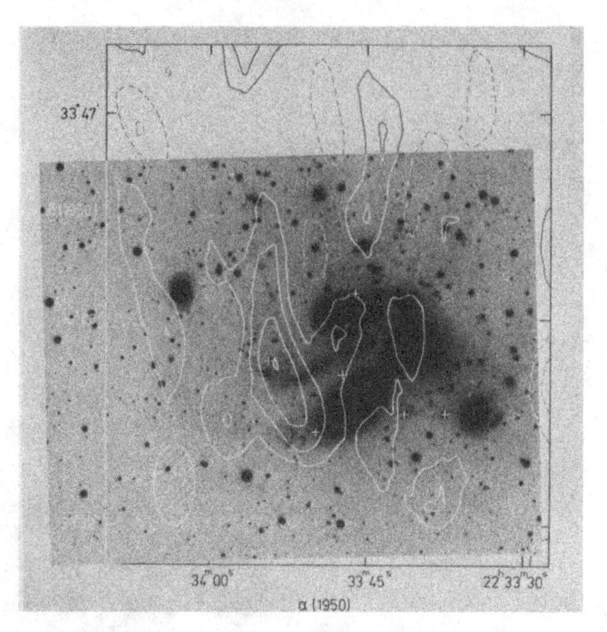

28 Map of a large cloud of HI in the velocity interval 6572 to 6633 km s^{-1} near Stephan's Quintet. From Allen and Sullivan (1980).

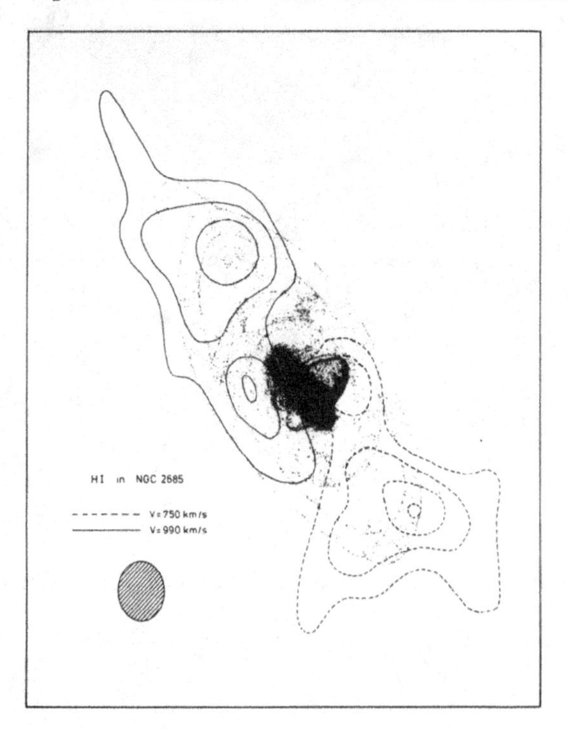

29 HI distribution at two extreme radial velocities in NGC2685. Described in Shane (1980).

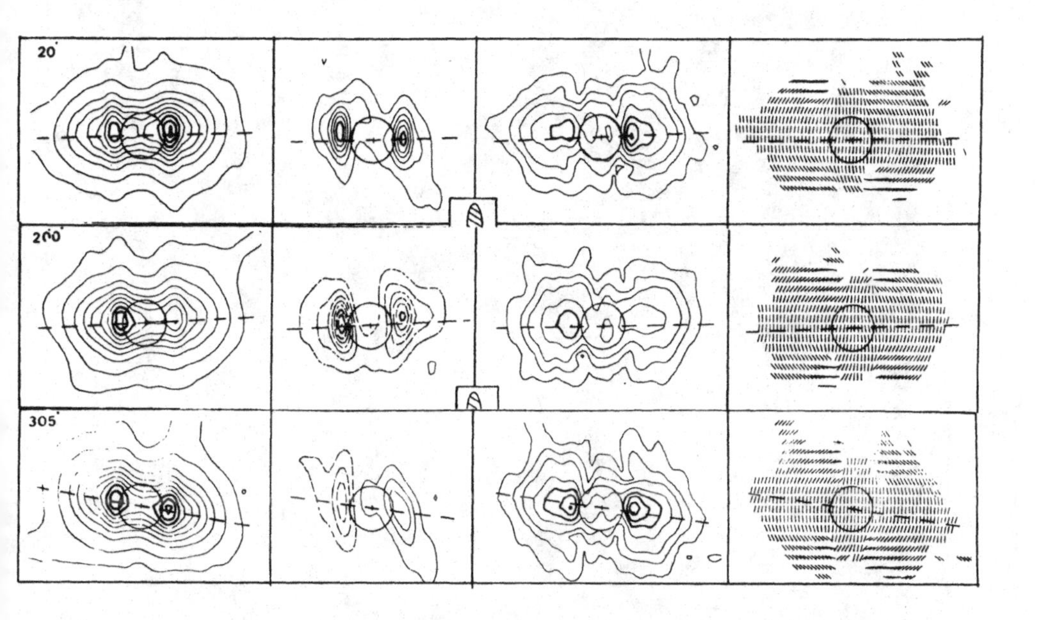

30 Maps of the continuum radio emission from Jupiter at 3 different longitudes. Panels from left to right show: intensity, circular polarization flux (left handed is dotted), linear polarization flux and magnetic field vectors. From De Pater (1980).

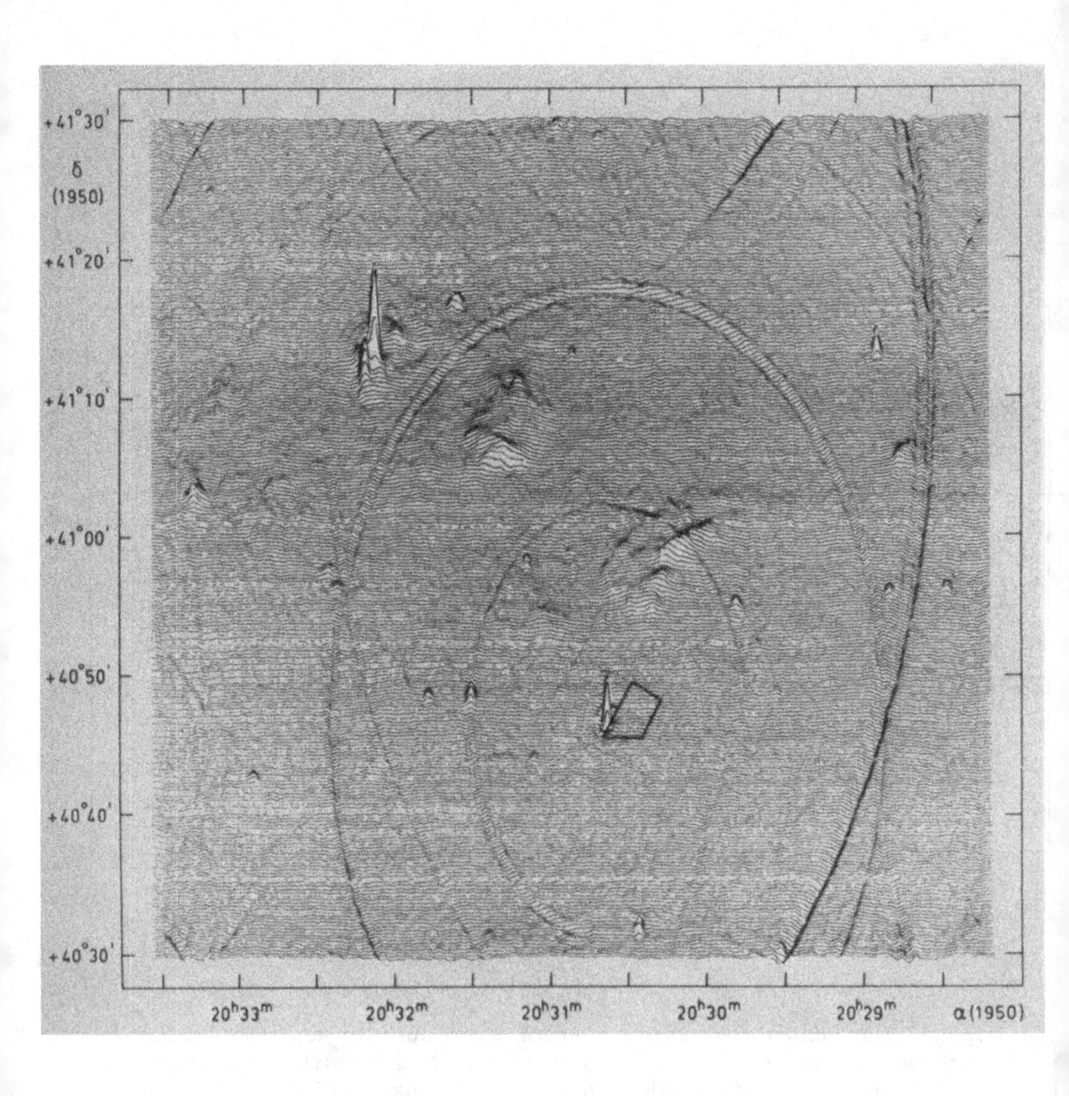

31 Map of the 21-cm continuum emission in the region of Cygnus X-3. From Braes and Miley (1972).

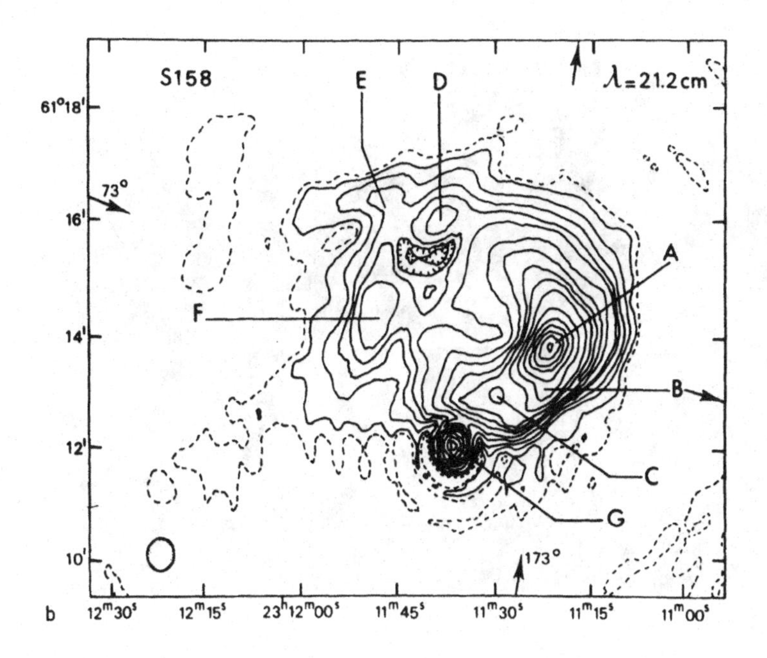

32 High-density condensations in NGC7538. From Israel, Habing and De Jong (1973).

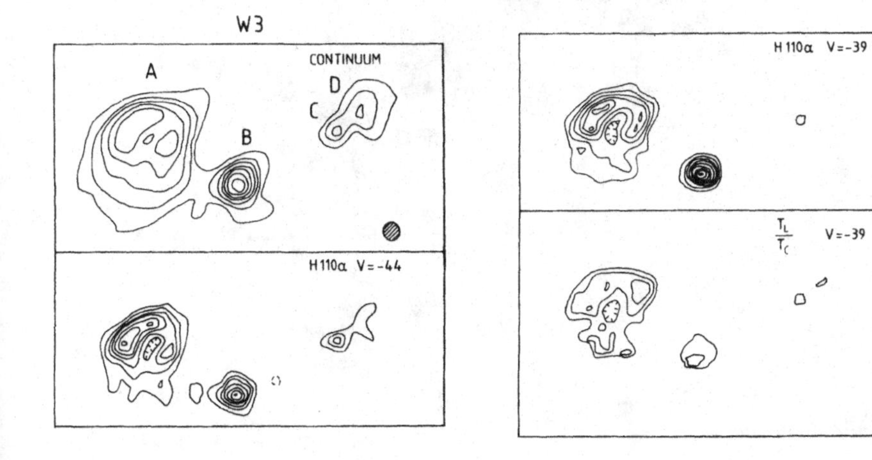

33 6-cm continuum and recombination line structure in W3. From Van Gorkom, Goss, Shaver and Harten (1980).

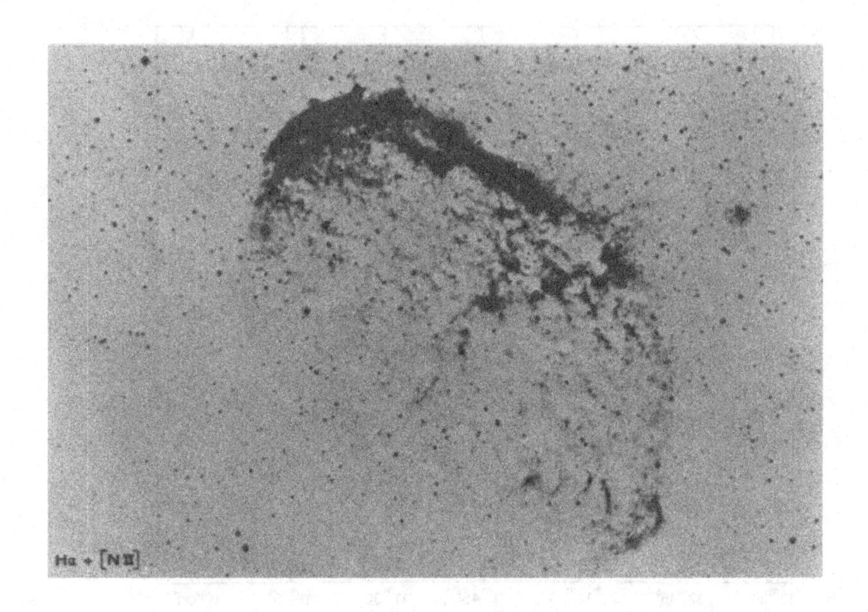

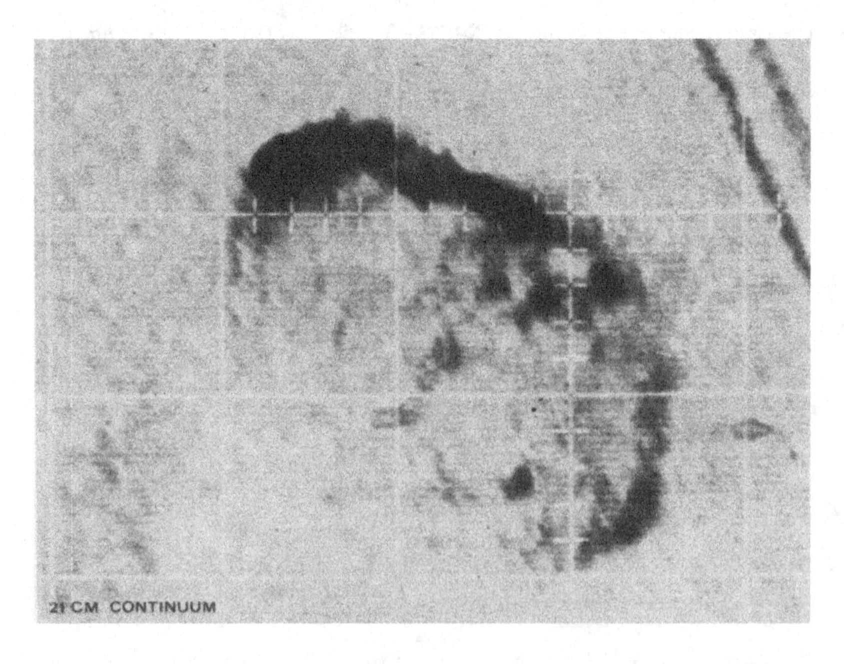

34 Optical emission lines and 21-cm radio continuum in NGC6888. From Wendker, Smith, Israel, Habing and Dickel (1975).

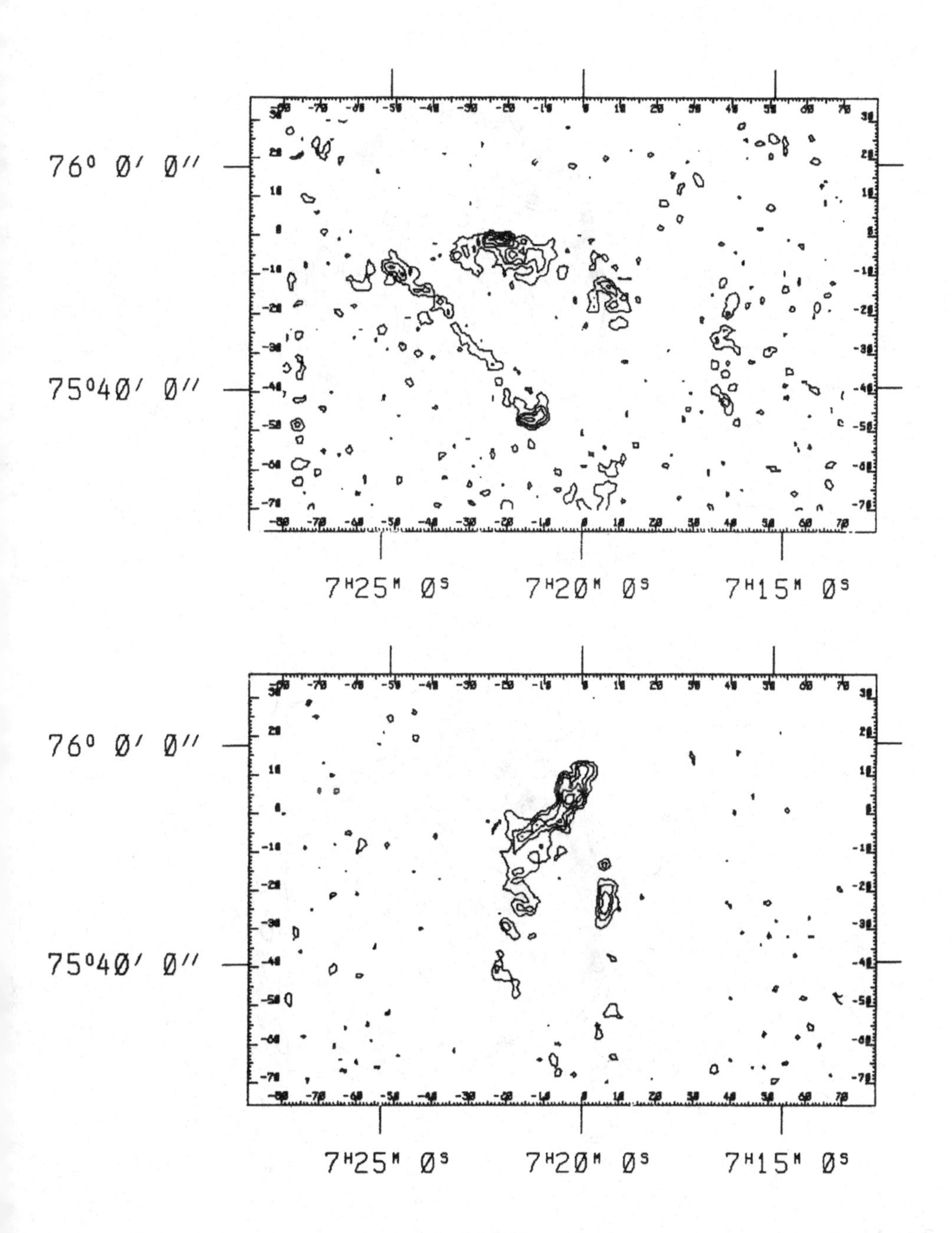

35 SRT spectrometer maps of HVC 139+28 at two radial velocities: -199 km s^{-1} (upper) and -203 km s^{-1} (lower). From work by Schwarz (private communication).

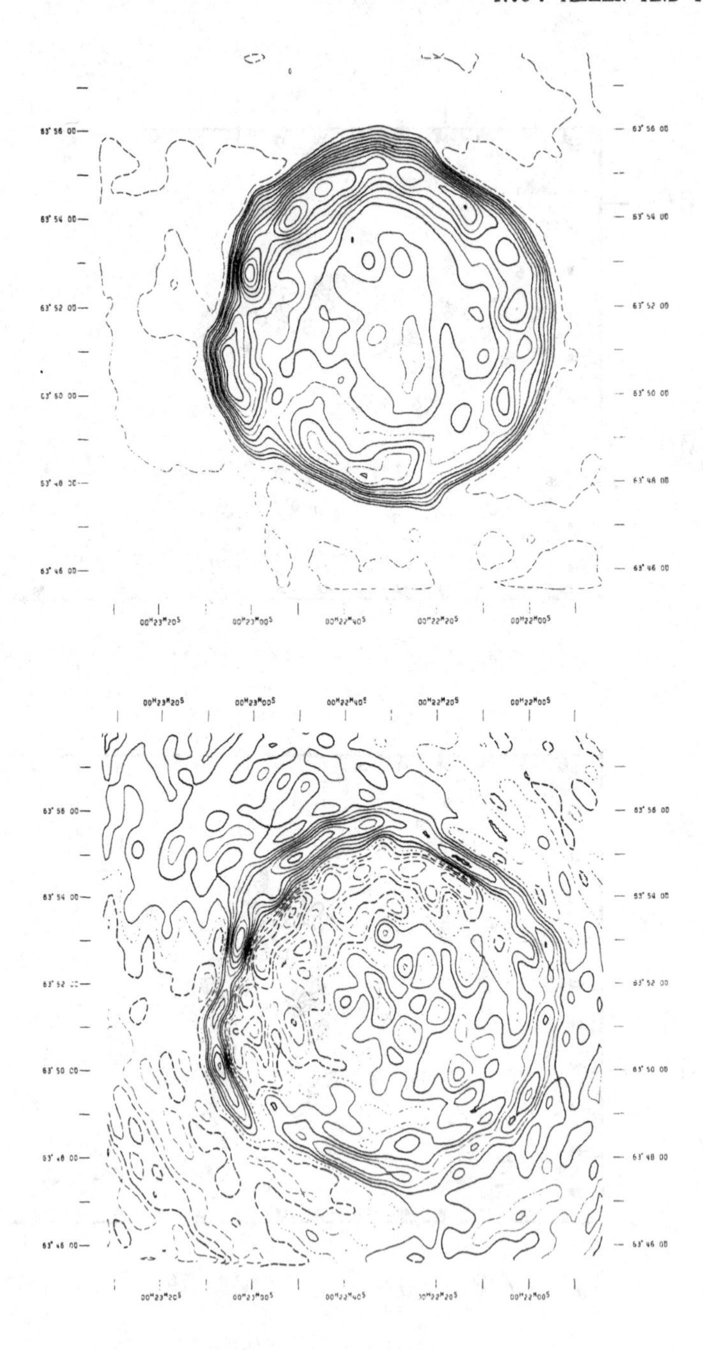

36 An epoch 1971 21-cm continuum observation of Tycho's supernova remnant (top) has been subtracted from a recent 1979 observation to show the results of the expansion (bottom). From Strom (private communication).

OORT'S WORK ON COMETS

Maarten Schmidt

This brief essay covers Oort's work on comets, in particular as I saw it in early 1950, during my first year at Leiden University as a graduate student. It was in the spring of 1949, at the annual astronomy conference in Doorn, that Jan Oort (actually: Professor Oort; it took almost twenty years before I addressed him as Jan) asked me to come to Leiden as "assistant". I arrived in the summer, after passing the "candidaatsexamen" at Groningen, and was immediately put to work by Professor Oosterhoff to estimate magnitudes of SV Centauri on plates taken with the Franklin-Adams telescope at Johannesburg.

Soon I was called for military duty, but was unexpectedly released after a couple of months. I returned to Leiden and was appointed again as assistant in November 1949. It was at that time that Oort suggested that I participate in investigating differences between old and new comets. This lasted till August 1950, when I left for Kenya to assist Van Herk at his temporary observatory at Timboroa in measuring declinations of fundamental stars. When I returned to Holland in the fall of 1951, the hydrogen 21-cm line had been detected and astronomy in Leiden had changed permanently.

Oort's work on the origin of comets was instigated by a study by A.J.J. van Woerkom (1948), published in B.A.N. No. 399. Van Woerkom showed that the passage of a comet through the inner parts of the solar system will typically change the inverse of the semi-major axis $\underline{a}$ of its orbit by $\pm$ 0.0005, mostly due to the perturbing effect of Jupiter. Hence, a long-period comet, coming in along an almost parabolic orbit, will after perihelion passage typically be in an elliptical orbit with a major axis of 4000 A.U., or escape from the solar system. Van Woerkom showed that, owing to Jupiter's perturbations, a field of parabolic comets will diffuse inward such that observed comets will have a uniform distribution over constant intervals of 1/a. The observed distribution of 1/a values, however, is not uniform, but shows a strong concentration at small positive values. Van Woerkom noted that the problem cannot be resolved by forces acting within the solar system.

The problems exposed by van Woerkom's work presented a challenge that, I imagine, Oort could not resist. Only a year after van Woerkom's paper, he published in B.A.N. No. 408 (Oort, 1950) his theory of "The

H. van Woerden, W. N. Brouw, and H. C. van de Hulst (eds.), Oort and the Universe, 111–115.

Structure of the Cloud of Comets Surrounding the Solar System, and a Hypothesis Concerning its Origin". Oort's picture of the origin and evolution of comets is ingenious and intellectually satisfying. I will attempt a brief description.

First, Oort emphasizes that comets must belong to the solar system. The incoming orbits of long-period comets show no well-established hyperbolic orbits. They do show substantial numbers of orbits with major axes in the range 20,000 to at least 150,000 A.U. Hence, there must exist a cloud of comets, reaching out to 150,000 A.U., that belongs to the solar system. Comets in the outer part share the Sun's velocity of 20 km/sec relative to nearby stars accurately, to within 30 cm/sec.

Next, Oort addresses the problem posed by van Woerkom's work, namely that Jupiter's action will remove from the cloud all long-period comets whose orbits enter the inner part of the solar system. He invokes the gravitational effect of nearby stars passing the solar system with its comet cloud to explain why we do observe long-period comets. The velocities of all comets in the cloud will be changed by each passing star. Even though these changes are small, they are sufficient to change the perihelion distance of a comet from, say, 10 A.U. (at which it would be unobservable) to one or two A.U., at which it would normally be detected. Oort shows that the gradual shuffling of velocities of comets in the cloud will insure a continuing supply of comets, passing through the inner part of the solar system, from distances of at least 50,000 A.U. From a detailed model of the cloud for which he works out the space distribution and the velocity distribution, he concludes that the cloud contains some 10^{11} comets.

As he stated in B.A.N. No. 408, Oort considered his hypothesis of the origin of the comet cloud "speculative". It does not seem particularly speculative at the present time, partly because it has proved to be a viable hypothesis, partly perhaps because astronomy as a whole has become more speculative. When the solar system was formed, a large number of small bodies were situated in orbits between Mars and Jupiter, many of which are now observed as minor planets or asteroids. Those with excentric orbits that carried them across Jupiter's orbit, must have had a close encounter with the planet at some time, which brought them into a long-period orbit. Subsequent perturbations by Jupiter will lead to further diffusion outwards, on the average. While many objects will escape from the solar system in the course of this process, a small fraction must have landed in very elongated orbits that carried them out to a distance between 50,000 and 200,000 A.U. These objects form the present comet cloud.

However, such a cloud of proto-comets would not be stable, since at each passage through the inner part of the solar system, the object would be subject to Jupiter's perturbation - effectively removing it from the outer part of the cloud. At this stage Oort invokes the gravitational effect of passing stars again, this time to diffuse the velocity components of the comets away from those that aim them at the inner part of the solar system. This increases their perihelion distance and safeguards the comets from Jupiter's strong perturbations. They are only subject to stellar perturbations; it is these that once in a while happen to re-aim a comet for the inner parts of the solar system, as discussed above. Such comets, when observed, may be called "new", since they have not been in the vicinity of the Sun since soon after the formation of the solar system.

The random nature of the perturbations by passing stars afforded a ready explanation of the random directions from which long-period comets are approaching the inner part of the solar system. With a probability of disintegration of 1 to 2 percent per perihelion passage, Oort could account for the distribution of the major axes larger than 50,000 A.U. He suspected that these "new" comets might exhibit different physical properties from those of "old" comets, which had been in the solar vicinity before, and it was at this stage that he involved me.

I spent several months early in 1950 collecting published data about observations of the brightness of comets. As usual, Oort showed continuing interest in the progress of the work. Every few weeks, at morning coffee in the "rekenkamer", he would - after consulting his notes on a small piece of paper or envelope - approach one and inquire how the work was progressing. Often he would ask one into his office and the discussion would continue till lunch time. At around 12:40 p.m. Mieke Oort would open the door of the hallway leading into their adjacent house and ring a bell to call Jan for lunch. He would usually not react, somewhat to my dismay, and sometimes Mieke would have to ring again before Jan would finally adjourn the discussion. I believe that Jan in later years became more prompt in this matter.

These discussions with Jan Oort had a special character. While one, at least initially, dreaded them, they always had a positive, inspiring effect. He was not didactic, yet enormously educational. Oort would not normally indicate how he reached a certain conclusion or point of view. Instead, he would present the arguments for his views with persuasive intellectual force. Those of his students who could follow the arguments benefited greatly.

The results of the work on differences between old and new comets were published in B.A.N. No. 419 (Oort and Schmidt, 1951). It turned out that new comets exhibit a much slower variation of brightness with heliocentric distance than do old comets. We argued that this difference could probably account for the excess of new comets observed over the numbers expected in a quasi-stationary state. On re-reading the article recently, I was struck by the discussion of selection effects, which was completed by Oort after I had left for Kenya in August 1950. This discussion is an essential part of the interpretation of the photometric results, to a degree that I only now appreciate.

The difference in photometric properties between old and new comets is not generally realized in the astronomical community. When the orbit of Comet Kohoutek became sufficiently well determined, it turned out to be highly elongated, allowing the possibility that it was in fact a new comet. If so, its predicted brightness at perihelion might be hundreds of times less than had been heralded by NASA publicity. I was pleased to read in a Dutch newspaper article that Jan had independently realized this. We were, it turned out, right.

In general, Oort's concept of the extended comet cloud and its origin has been well received in the astronomical community. The main opposition has come from R.A. Lyttleton, who proposes to explain comets as swarms of small particles captured by the Sun in the wake of past slow passages through interstellar clouds. Lyttleton has argued in particular against Oort's interpretation of the distribution of the inverse semi-major axis 1/a. He states that Oort's deduction, that there must be a maximum in the distribution of semi-major axes a because the observed distribution of 1/a shows a maximum, is incorrect. This sounds impressive until one realizes that Oort does not refer to such a maximum at all. All Oort does assert is that many of the observed comets come from distances around 100,000 A.U. One sees quite a few cases where controversy in astronomy is generated by arguments aimed at disproving statements that were never made. Oort does not like to debate controversial issues and has not, I believe, responded to Lyttleton's arguments. R.S. le Poole and P. Katgert (1968) later explained the situation in a letter to Observatory magazine. In discussing other objections raised by Lyttleton, they present a nice compact summary or Oort's theory of the origin and evolution of the extended cloud of comets.

Even though comets were a relatively minor interest of Jan Oort's - he did not mention them in his "afscheidscollege" (retirement lecture) in 1970 -, his contribution was similar in character to those he had given in many other fields: creative, not needlessly speculative, thorough, and important. His papers are a joy to read, as I noticed in particular on re-reading his Halley Lecture (Oort, 1951). He does, of course, take great care in preparing articles and even then makes numerous changes and corrections in galley proofs, and in earlier days in page proofs, no doubt to editors' distress.

While his articles are a lasting record of his varied contributions to astronomy, it is Jan Oort's personal qualities that are most impressive. Those of us who know him well are privileged by the experience. Jan's friendship is my treasure. Happy birthday, Jan!

References

Le Poole, R.S. and Katgert, P. 1968, Observatory 88, pp. 164.
Oort, J.H. 1950, Bull. Astr. Inst. Netherlands 11, pp. 91 (No. 408).
Oort, J.H. 1951, Observatory 71, pp. 129.
Oort, J.H. and Schmidt, M. 1951, Bull. Astr. Inst. Netherlands 11, pp. 259 (No. 419).
Van Woerkom, A.J.J. 1948, Bull. Astr. Inst. Netherlands 10, pp. 445 (No. 399).

Maarten Schmidt studied at Groningen and Leiden, 1946-1956, and has since been at the Mount Wilson and Palomar Observatories.

Maarten Schmidt

During a semaine d'étude at the Vatican: Schmidt, McCarthy, Woltjer, Oort and Fowler.

THE EVOLUTION OF IDEAS ON THE CRAB NEBULA

L. Woltjer

The Crab Nebula is believed to have been discovered by John Bevis. It was first catalogued by Messier (M 1). It was named and described by the Earl of Rosse, who observed it in the middle of the last century with his six-foot telescope in Ireland. Extensive studies by Lampland (1921a) at Flagstaff rather convincingly revealed changes in the Nebula - although the difficulty in observations of this type is put in evidence by his finding (Lampland 1921b) of variations in the Sc galaxy NGC 4254 which appear not to have been confirmed. Slipher (1916) and Sanford (1919) obtained spectra, which revealed a continuum with superimposed emission lines split into two components; this reminded Slipher of the Stark effect, but was later interpreted as due to expansion. Lundmark (1921) in a list of suspected novae includes the "guest star" of 1054 from the list by Ma-Tuan-Lin translated by Biot (1846), and makes the brief comment "near NGC 1952" (the Crab Nebula).

The evidence for variations in the Nebula was confirmed by Duncan (1921), who found that it had expanded (in radius) by about 1.5 arc-seconds in eleven years, corresponding to an age of the order of 900 years (see further Duncan 1939), and Hubble (1928) noted that this made the identification with the star of 1054 more probable. Further confirmation came from the work of Deutsch and Lavdovsky (1940), who independently measured the expansion age and found the expansion to have started in 1155 ± 140 years.

New spectra were obtained with the nebular spectrograph at the Crossley reflector at Lick by Mayall (1937). These spectra clearly showed that the split emission lines merge near the edge of the Nebula, exactly as would be expected in an expanding shell. From a comparison of the astrometric expansion data with the radial velocities, the distance may be obtained. Following Lundmark's (1926) first application of this method, Mayall on the basis of his own radial velocities and Duncan's (1921) expansion data found a distance of 1500 pc. Subsequent estimates based on different data sets and different assumptions about the nebular geometry have generally all been in the range 1500 ± 500 pc.

By the end of the thirties, various lines of evidence therefore supported the identification of the Crab Nebula and the 1054 event, but the information about this event was still very limited. It is from this

H. van Woerden, W. N. Brouw, and H. C. van de Hulst (eds.), Oort and the Universe, 117–122.

time that Oort's first work on the Crab Nebula dates. As a result of a collaboration with Duyvendak, professor of oriental languages at Leiden, and with Mayall at Lick, two companion papers were published. In the first, by Duyvendak (1942), several Chinese texts are collected and translated which give information about the light curve of the guest star of 1054. In the second paper, Mayall and Oort (1942) studied the astronomical implications, established that the guest star must have been a supernova, and concluded that "The Crab Nebula may be identified with the 1054 supernova".

In the meantime, at Mount Wilson, Zwicky, Baade and Minkowski had started their studies of supernovae. Several light curves were obtained by Baade, while on the basis of his extensive spectroscopic studies, Minkowski (1941) was able to recognize two distinct classes of supernovae, "provisionally" called type I and type II. While the latter showed some spectroscopic similarity to the novae, the former had spectra with broad unidentified emission bands. The spectra and light-curves of type I supernovae showed a remarkable uniformity. The best observed case, the supernova of 1938 in IC 4182, was observed spectroscopically till 339 days and photometrically till 635 days after maximum. (It is hard to see how at present, with observers at large telescopes generally receiving allocations of a few nights for predictable projects, studies like those of Baade and Minkowski still could be executed!) On the basis of these studies and the data discussed by Mayall and Oort, Baade (1942) concluded that the guest star of 1054 had been a supernova of type I. Later, however, Minkowski (1968) expressed some doubts.

The first phase of Crab Nebula research was completed by two papers by Baade (1942) and Minkowski (1942). Baade reanalyzed the astrometric data of Duncan (1939) and concluded that the expansion age of the Nebula (derived on the basis of a constant expansion speed) was 130 $\pm$ 40 years less then the real age. Baade inferred from this that the Nebula was in a state of accelerated expansion. Since he was unable to find a reasonable physical mechanism that could quantitatively account for the acceleration, he finally concluded that there was "only one way out of these difficulties, namely to assume that the measured rate of the angular expansion is too large and that the resulting acceleration of the expansion is spurious". Later studies (Trimble 1968), however, have fully confirmed the acceleration, and Pikelner (1956) was the first to point out that the pressure of the magnetic fields and relativistic particles on the nebular shell adequately explains the acceleration. On the basis of the astrometric and spectroscopic data, Baade and Minkowski also (correctly) identified one of the stars near the centre of the Nebula as the stellar remnant of the supernova. Perhaps the most important result of their investigations, however, was the photometric and spectroscopic evidence they presented that the Nebula consists of two distinct parts: an outer system of filaments with an emission-line spectrum and an inner mass of more amorphous structure with a continuous spectrum. The interpretation of the continuous spectrum led to some difficulties: Minkowski fitted it by Bremsstrahlung from a hot massive cloud of gas, energized by the central star. The relation between this gas and the cool, low-mass filaments was far from obvious.

The nature of the Cygnus Loop was investigated by Oort (1946) in his Darwin lecture, with the conclusion that this was most probably the shell of a supernova slowed down by interaction with interstellar matter. In fact, it now appears that most known supernova remnants result from this process. Oort derived the fundamental parameters of the supernova shell on the basis of the assumption that it was in the momentum-conserving phase; he so obtained entirely reasonable values. In a subsequent investigation, Oort (1951) pointed out some of the difficulties facing current models of the Crab Nebula, and showed that it might in the future evolve into an object like the Cygnus Loop and that the interaction with interstellar matter might be important already now in the energetics of the Crab Nebula.

The problems relating to the continuum were further aggravated by the discovery of radio emission from the Crab Nebula (Bolton and Stanley 1949) with a flux that was impossible to explain on the basis of the models advanced for the optical continuum. An entirely new picture was called for.

The solution to these problems came from studies by Alfvén and Herlofson (1950) and by Kiepenheuer (1950), who pointed out the potential importance of synchrotron radiation (due to relativistic electrons in a magnetic field) in the generation of radio waves. In particular, Kiepenheuer showed that on the basis of this mechanism the galactic non-thermal radio emission could be understood with entirely reasonable values for the density of cosmic-ray electrons and for the strength of the galactic magnetic field. These ideas and their connection with cosmic-ray physics were further pursued by Ginzburg (1951) and his associates. Subsequently, Shklovsky (1953) indicated their importance to the resolution of the difficulties in the Crab Nebula, proposing that both the radio and the optical continuum emission derive from this mechanism, and showing that then the relative intensities of the optical and radio emission could be naturally accounted for. Gordon (1954) and Ginzburg (1954) then predicted that the continuum of the Crab should be linearly polarized. The same year, Dombrovsky (1954) reported the first optical observations of polarization in the Crab Nebula, which were confirmed by Vashakidze (1954). Dombrovsky observed five regions in the Crab with the 40-cm telescope at Bjurakan and found polarizations ranging from 9 to 15%. More detailed observations were made by Khachikyan (1955), who found that in some parts of the Nebula the polarization reached values in excess of 50%.

In the meantime, Oort and Walraven had taken up an observational study of the Crab Nebula. An instrument was constructed for the measurement of the integrated brightness of the Nebula to look for possible time variations. Upon learning of the observations in the USSR, they quickly converted this instrument into a polarimeter. The heart of the instrument consisted of a special 19-stage photomultiplier, made by Lallemand at Paris Observatory. With this device, they established with good accuracy the integrated magnitude and polarization of the Nebula. In addition, a set of more than 50 detailed polarization observations were made in various spots, with diaphragms as small as 30 arcseconds.

It is still remarkable that such accurate observations could be made with a 13-inch telescope in Leiden, where the climate and the city lights create the most infavourable circumstances. These observations reliably outlined the basic magnetic-field structure in the Nebula. In their study (Oort and Walraven 1956) several other important points were made:

(1) The distribution of the optical emission is much narrower than that of the radio emission, a fact which is of much importance in the analysis of the distribution of relativistic electrons and magnetic fields in the Nebula.

(2) The ultraviolet extension of the synchrotron spectrum of the Nebula may suffice to ionize the filaments.

(3) From a comparison of photographs taken at various epochs (including one at Lick in 1899), the variations observed by Lampland are amply confirmed and appear to be occurring in the continuum. From these observations and those of Baade in the central regions, it is clear that perturbations arise there and then propagate further out.

As usual, Oort's work inspired others to follow: Walraven (1957) obtained more polarization observations with an 80-cm telescope at Haute Provence. Woltjer (1957) did photometry on Baade's (1956) plates, which yielded a very detailed picture of the distribution of intensity and polarization. Later, in a thesis under Oort's direction (Woltjer 1958), spectrophotometric measurements were made on Mayall's plates, and a detailed analysis of the conditions in the filaments as well as a discussion of the dynamics of the Nebula were given. Westerhout attempted to measure the radio polarization of the Nebula at 21 cm with a 7.5-meter radio telescope, and Seeger and Westerhout (1957) observed the 1955 lunar occultations at Dwingeloo and obtained a more accurate picture of the distribution of the radio radiation.

At the end of this period, most aspects of the optical and radio radiation from the Crab Nebula had been considered. Two fundamental discoveries remained to be made: The Crab Nebula was discovered to be an X-ray source by Gursky, Giacconi, Paolini and Rossi (1963) and by Bowyer, Byram, Chubb and Friedman (1964a). The latter authors (1964b) later showed in a lunar-occultation experiment that the source was extended. Shklovsky (1964) and Woltjer (1964) demonstrated that a synchrotron-radiation interpretation of the X-rays was plausible. This was confirmed by the detection of X-ray polarization by Novick, Weisskopf, Berthelsdorf, Linke and Wolff (1972). Perhaps even more important was the discovery that the star which Baade and Minkowski with remarkable foresight had singled out as the stellar remnant of the supernova, was in fact a pulsar at radio wavelengths (Staelin and Reifenstein 1968), in the optical (Cocke, Disney and Taylor 1969), and in X-rays (Fritz, Henry, Meekins, Chubb and Friedman 1969).

The Crab Nebula has played a fundamental role in several areas of astrophysics. It has shown the importance of relativistic particles and magnetic fields, thus creating a direct link between astronomical observations and cosmic-ray studies. It has given the conclusive proof that pulsars are rotating neutron stars which with high efficiency generate energetic particles. While the Nebula is much better understood now than it was when Oort became first interested in it, major problems remain to be solved in the future. For example, the chemical composition of the shell is only partly known and its connection with nucleogenesis is still unclear. Also, a full understanding of particle propagation and acceleration in the Nebula will be much advanced by observations of the spatial distribution of hard X-rays.

The Crab Nebula has not been the main line of Oort's research, but it has been an object that has interested him over several decades and to which he has returned repeatedly. Oort's impact on the evolution of our knowledge of the Crab Nebula has been substantial, both through his own researches and through his influence on those who were fortunate enough to be his coworkers. It now seems clear that in many ways the Crab Nebula may be a prototype for the much more powerful events which occur in active galaxies, a subject which has such an important place in Oort's current researches.

References

Alfvén, H. and Herlofson, N. 1950, Phys. Rev. 78, pp. 616.
Baade, W. 1942, Astrophys. J. 96, pp. 188.
Baade, W. 1956, Bull. Astr. Inst. Netherlands 12, pp. 312.
Biot, E. 1846, Connaissance des Temps.
Bolton, J.G. and Stanley, G.J. 1949, Austr. J. Sci. Res. 2A, pp. 139.
Bowyer, S., Byram, E.T., Chubb, T.A. and Friedman, H. 1964a, Nature 201, pp. 1307.
Bowyer, S., Byram, E.T., Chubb, T.A. and Friedman, H. 1964b, Science 146, pp. 912.
Cocke, W.J., Disney, M.J. and Taylor, D.J. 1969, Nature 221, pp. 525.
Deutsch, A.N. and Lavdovsky, V.V. 1940, Pulkovo Circ. 30.
Dombrovsky, V.A. 1954, Doklady Akad. Nauk SSSR 94, pp. 1021.
Duncan, J.C. 1921, Mt. Wilson Communication 76.
Duncan, J.C. 1939, Astrophys. J. 89, pp. 347.
Duyvendak, J.J.L. 1942, Publ. Astr. Soc. Pacific 54, pp. 91.
Fritz, G., Henry, R.C., Meekins, J.F., Chubb, T.A. and Friedman, H. 1969, Science 164, pp. 709.
Ginzburg, V.L. 1951, Doklady Adak. Nauk SSSR 54, pp. 91
Ginzburg, V.L. 1954, Trudy Sov. Vopr. Kosmog. 3, pp. 258
Gordon, I.M. 1954, Doklady Akad. Nauk SSSR 94, pp. 413.
Gursky, H., Giacconi, R., Paolini, F.R. and Rossi, B.B. 1963, Phys. Rev. Lett. 11, pp. 530.
Hubble, E. 1928, Astr. Soc. Pacific Leaflet 14.
Khachikyan, E.E. 1955, Doklady Akad. Armenian SSSR 21, pp. 63.
Kiepenheuer, K.O. 1950, Phys. Rev. 79, pp. 738.
Lampland, C.O. 1921a, Publ. Astr. Soc. Pacific 33, pp. 79.
Lampland, C.O. 1921b, Publ. Astr. Soc. Pacific 33, pp. 167.
Lundmark, K. 1921, Publ. Astr. Soc. Pacific 33, pp. 225.
Lundmark, K. 1926, Uppsala Astr. Obs. Medd. 12.
Mayall, N.U. 1937, Publ. Astr. Soc. Pacific 49, pp. 101.
Mayall, N.U. and Oort, J.H. 1942, Publ. Astr. Soc. Pacific 54, pp. 95.
Minkowski, R. 1941, Publ. Astr. Soc. Pacific 53, pp. 224.
Minkowski, R. 1942, Astrophys. J. 96, pp. 199.
Minkowski, R. 1968, Stars and Stellar Systems, 7, ch. II, (Univ. of Chicago Press).

Novick, R., Weisskopf, M.C., Berthelsdorf, R., Linke, R. and Wolff, R.S. 1972, Astrophys. J. Lett. 174, pp. L1.
Oort, J.H. 1946, Mon. Not. Roy. Astr. Soc. 106, pp. 159.
Oort, J.H. 1951, Problems of Cosmical Aerodynamics (Central Air Documents Office, Dayton, Ohio) pp. 118.
Oort, J.H. and Walraven, Th. 1956, Bull. Astr. Inst. Netherlands 12, pp. 285.
Pikelner, S.B. 1956, Astr. Zhurn. USSR 33, pp. 785.
Sanford, R.F. 1919, Publ. Astr. Soc. Pacific 31, pp. 108.
Seeger, Ch.L. and Westerhout, G. 1957, Bull. Astr. Inst. Netherlands 13, pp. 312.
Shklovsky, I.S. 1953, Doklady Akad. Nauk SSSR 90, pp. 983.
Shklovsky, I.S. 1964, Astron. Tsirk. 304.
Slipher, V.M. 1916, Publ. Astr. Soc. Pacific 28, pp. 192.
Staelin, D.H. and Reifenstein, E.C. 1968, Science 162, pp. 1481.
Trimble, V. 1968, Astron. J. 73, pp. 535.
Vashakidze, M.A. 1954, Astron. Tsirk. 147, pp. 11.
Walraven, Th. 1957, Bull. Astr. Inst. Netherlands 13, pp. 293.
Westerhout, G. 1956, Bull. Astr. Inst. Netherlands 12, pp. 309.
Woltjer, L. 1957, Bull. Astr. Inst. Netherlands 13, pp. 301.
Woltjer, L. 1958, Bull. Astr. Inst. Netherlands 14, pp. 39.
Woltjer, L. 1964, Astrophys. J. 140, pp. 1309.

Lodewijk Woltjer studied at Leiden in 1948-1957. He was a professor there from 1960 to 1964 , and then became professor at Columbia University in New York. Since 1975 Dr. Woltjer is Director-General of the European Southern Observatory.

OORT'S WORK REFLECTED IN CURRENT STUDIES OF GALACTIC CO

W.B. Burton

When I was a graduate student in Leiden during the 1960's, several areas of research in galactic structure were of vital interest to Professor Oort. These areas included the dynamical influence of the unseen mass, the enigmatic high-velocity clouds, the structure and influence of the galactic nucleus, and the form of spiral structure in the Galaxy. Each of these areas of research flourishes now in ways which could not have been foreseen. That they flourish indicates the very special nature of Professor Oort's insight and taste in problems, and the encouragement and inspiration which he gives to others. My comments here are limited to brief remarks on current investigations of CO in the Galaxy with which I am involved. The topics represented in current CO research focus on several areas whose foundations were originally motivated to a large extent by Professor Oort.

Regarding the unseen mass:

The mass density near the Sun necessary to provide the measured component of the galactic gravitational field perpendicular to the galactic plane is not accounted for by the observed constituents of the Galaxy (Oort 1960, also Oort 1932). Oort (1965) considered several forms of matter in which the unseen mass might reside. One of these was molecular hydrogen. Although it had been thought for some time (see e.g. Oort 1946) that hydrogen in molecular form would be an abundant constituent of the interstellar medium, the absence of a widely observable transition in molecular hydrogen precluded testing the hypothesis in a direct way until the recent observations of the tracer molecule CO became available. Before the direct evidence, however, Oort (1965) had been able to reject H_2 as the principal constituent of the unseen mass. If H_2 were to contribute the unseen mass, Oort argued, it would require some six times more hydrogen in molecular form than in the atomic form. Kootwijk and Dwingeloo results had shown that atomic-hydrogen surface densities increase outward with distance from the Sun. Oort argued that if the total gas followed the HI increase, then, in order for the rotation curve to drop in the way supposed at that time, the stellar mass would have to fall off with increasing distance much more abruptly than followed by observation. The information revealed

H. van Woerden, W. N. Brouw, and H. C. van de Hulst (eds.), Oort and the Universe, 123–128.

by the tracer CO molecule since Oort's discussion shows in a rather direct way that molecular hydrogen is not present in sufficient quantities to supply the unseen mass. Although most of the unseen mass in the inner Galaxy is in the H_2 form, one of the most surprising results of the CO surveys is the discovery that the H_2 density does not follow the atomic-hydrogen density. In the outer Galaxy the atomic hydrogen dominates, but in the inner galaxy H_2 dominates the interstellar medium. The problem of the unseen mass remains, but at an even increased level of interest. Not only must we seek for the missing mass necessary to provide the observed dynamical situation perpendicular to the equator, but we must in addition find the yet greater quantities of material sufficient to keep the rotation curve in the outer galaxy flat. Clearly we have now no clear indication of the state in which this mass is hidden. The total quantity of unseen mass will be given by the galactic rotation parameters which are currently being subjected to intensive study. The foundations of this continuing study were, of course, laid by Professor Oort (e.g. 1927).

Regarding the high-velocity clouds:

One of the most surprising results to emerge from the Dwingeloo telescope in the early 1960's was the discovery of high-velocity clouds of atomic hydrogen. An extensive series of observations made by astronomers from Leiden and Groningen showed that large arcs on the sky, especially in the second and third quadrants of galactic longitude, were described by hydrogen at negative velocities with amplitudes some 100 or 200 km s^{-1} greater than permitted by circular galactic rotation. These clouds remain enigmatic, largely because their distances are unknown, and because they have never been detected except in the HI line. Oort (1967) reviewed the plausible interpretations for these clouds, and favoured interpreting them as material in a quite primordial state outside the Galaxy proper and therefore capable of revealing much about the formation and early history of the Galaxy (Oort 1969, 1970). The recent discoveries of extended hydrogen emission around many external galaxies, in some cases bridging the space amongst galaxies in close groups, support the early suggestion of Professor Oort. Also supporting his interpretation that the material in the high-velocity clouds is generally unprocessed, is the lack so far of a detection in these clouds of material other than HI. Many groups have searched for CO in high-velocity clouds. CO serves as a good tracer of the physical conditions in the interstellar medium. Its presence in observable quantities requires shielding from photodissociation, low temperatures, and high densities of gas and dust grains. These conditions are evidently not met in high-velocity clouds to the extent they are, for example, in diffuse dark clouds in the galactic layer. Most recently Liszt and I searched for CO in positions given to us by Professor Oort from Westerbork HI observations which had shown knots of high HI column densities with length scales of some minutes of arc. These positions provided apparently good candidates for CO observations, but after lengthy integrations we were nevertheless unable to detect CO emission in these knots. Although without knowledge of the excitation temperature in these clouds it is difficult to derive limits on the molecular column density, it seems nevertheless to be the case that the high-velocity clouds are not the sites of large quantities of processed material.

Regarding the structure and influence of the galactic nucleus: It seems to me that hindsight allows the assertion that the most important results derived during the 1960's at Leiden from the 21-cm line are those concerning the structure of the nucleus of our Galaxy. The theses written under Professor Oort's supervision by G.W. Rougoor and P.C. van der Kruit showed a variety of features deviating strongly from circular motions and containing substantial amounts of material (see Oort and Rougoor, 1960). Although the 3-kpc arm and the other anomalous features in the galactic nucleus still lack convincing dynamical interpretation, recent work shows the relevance of these observations to the subject of galaxies in general. Westerbork observations have revealed an analog to the 3-kpc arm in the Andromeda Galaxy, and many galaxies show nuclear characteristics not very different from our own (Oort, 1971). Oort has persistently suggested that phenomena occurring in galactic nuclei may be responsible for many aspects of a galaxy's appearance (see e.g. Van der Kruit, Oort, and Mathewson, 1972). This persistence will undoubtedly be rewarded, because the study of the general phenomena of galactic nuclei is now one of the most vital areas of astronomical research. Observations of CO are playing an increasingly important role in these studies (Oort, 1977). In our own and in other galaxies the short wavelength of the CO transition allows high resolution, the density of nuclear gas provides strong signals, the rapidly varying innergalaxy velocity fields provide much kinematic information, and the low temperature of the molecular gas presents this kinematic information at high resolution. CO observations made by T.M. Bania in our own Galaxy give the gaseous mass of the 3-kpc arm, and, because there is less intervening diffuse emission in the CO case than in the HI case, allow the 3-kpc arm to be followed over a very large arc. Several groups have been studying the detailed structure in the nucleus: these efforts will continue for some time, because extensive data can only be accumulated slowly with a 1-minute beam. Our results show spatial and kinematic regularities in the inner-galaxy CO emission which are consistent with an interpretation in terms of a coherent, symmetric body of gas centred on the nucleus. Consequently we doubt the need for a variety of separately ejected gas clouds. The CO data support the conclusion drawn from HI observations in the nucleus by several groups that the overall gas distribution is tilted with respect to the galactic equator. Dynamical interpretation of this tilt has yet to be given: likewise the dynamics of the 3-kpc arm and other nuclear features in clearly non-circular motion are not yet understood.

Regarding the form of the spiral structure in the Galaxy: Some of the motivation for the early HI surveys centred on the hope that the transgalactic paths accessible at 21-cm wavelength together with the kinematic information available in the spectral line would allow mapping of the spiral structure of the Galaxy. Although of course the HI line has provided an enormous amount of information on galactic morphology (e.g. Oort, Kerr, and Westerhout, 1958) and on the physical state of the interstellar medium, the particular hope for a detailed map of the Galaxy has in some respects been frustrated. Much of this frustration has been attributed to the rather smooth, ubiquitous distribution of the HI gas, which renders the resulting spectra very susceptible to kinematic perturbations, and to the rather low density contrast (~3:1) predicted between eventual arm- and interarm regions. Some of the motivation for carrying out the large-scale surveys for CO likewise focused on the hope of mapping the spiral structure. In the case of the CO tracer, this hope was reinforced by a range of results reached by the mid-1970's. Comparison of optical studies with aperture-synthesis radio data showed that the narrow dust lanes in external galaxies portray spiral arms in a much more definite way than do HI maps (see Oort, 1975). Theoretical investigations had confirmed that the diffuse gas would probably respond in a linear manner to the passage of the density wave, whereas the cold, dense material would experience a non-linear response along a very narrow shock front. This narrow shock front might indeed precipitate the formation of the CO clouds. In addition, the early CO data showed that the clouds occupied only about 1% of the total volume of the galactic layer, leading to a tendency to consider unimportant the velocity-crowding and blending effects which dominate the spectra from HI (which occupies essentially 100% of the galactic-layer volume). Liszt and I have recently considered the validity of ignoring these effects for the CO data. We did this by generating synthetic spectra corresponding to a model of the molecular-clouds ensemble derived in accordance with the CO-survey results. We concluded that the transformation from the galactic spatial and kinematic coordinates to the observed position-radial velocity maps involved for the CO data the same sort of consequences that have frustrated galactic mapping efforts based on HI data. This conclusion holds, despite the very small volume-filling factor of the ensemble of individually opaque, discrete clouds. When the characteristics of the actual observing procedure and of galactic kinematics are accounted for, radiative transfer through the ensemble is seen to result in a macroscopically optically thin line; consequently the influence of blending of individual CO features occurs in a way analogous to the velocity-crowding dominating the 21-cm profiles. Because much structure in the CO spectra results primarily from the characteristics of the galactic velocity field and only secondarily from localized density variations, we find that the evidence for spiral structure in the currently available CO surveys is still rather elusive. This modelling procedure has, however, given us parameters for the molecular-cloud ensemble which we consider reliable. We are using these parameters now to consider the dynamics of the thickness of the cloud layer, which is greater than allowed by the individual random motions of the clouds (Oort, 1965), and the general problem of the dynamical equilibrium between the cloud ensemble and the stars (Spitzer and Schwarzschild, but see Oort and Spitzer, 1955).

Acknowledgements:

My work on galactic structure from CO data is being done in collaboration with Harvey S. Liszt of the National Radio Astronomy Observatory, and is supported by the National Science Foundation through grant AST-7921812.

References

Oort, J.H. 1927, "Investigations Concerning the Rotational Motion of the Galactic System, Together with New Determinations of Secular Parallaxes, Precession and Motion of the Equinox," Bull. Astr. Inst. Netherlands 4, pp. 79.
Oort, J.H. 1932, "The Force Exerted by the Stellar System in the Direction Perpendicular to the Galactic Plane, and Some Related Problems," Bull. Astr. Inst. Netherlands 6, pp. 249.
Oort, J.H. 1946, "Some Phenomena Connected with the Interstellar Matter," Mon. Not. Roy. Astron. Soc. 106, pp. 159 (The George Darwin Lecture, 1946).
Oort, J.H. 1952, "Problems of Galactic Structure," Astrophys. J. 116, pp. 233 (The Henry Norris Russell Lecture, 1951).
Oort, J.H. 1960, "Note on the Determination of K_z and on the Mass Density Near the Sun," Bull. Astr. Inst. Netherlands 15, pp. 45.
Oort, J.H. 1965, "Stellar Dynamics," in "Galactic Structure", eds. A. Blaauw and M. Schmidt, "Stars and Stellar Systems," V, pp. 455, Univ. Chicago Press.
Oort, J.H. 1967, "Possible Interpretations of the High-Velocity Gas," in "Radio Astronomy and the Galactic System," ed. H. van Woerden, I.A.U. Symp. 31, pp. 279.
Oort, J.H. 1969, "Infall of Gas from Intergalactic Space," Nature 224, pp. 1158.
Oort, J.H. 1970, "The Formation of Galaxies and the Origin of the High-Velocity Hydrogen," Astron. Astrophys. 7, pp. 381.
Oort, J.H. 1971, "Composition and Activity of the Nucleus of our Galaxy, and Comparison with M31", in "Nuclei of Galaxies", ed. D.J.K. O'Connell, pp. 321, North-Holland Publ., Amsterdam.
Oort, J.H. 1975, "Phenomenology of Spiral Galaxies," in "Structure and Evolution of Galaxies," ed. G. Setti, pp. 85, Reidel, Dordrecht.
Oort, J.H. 1977, "The Galactic Center," Ann. Rev. Astron. Astrophys. 15, pp. 295.
Oort, J.H., Kerr, F.J. and Westerhout, G. 1958, "The Galactic System as a Spiral Nebula," Mon. Not. Roy. Astron. Soc. 118, pp. 379.
Oort, J.H. and Rougoor, G.W. 1960, "Distribution and Motion of Interstellar Hydrogen in the Galactic System with Particular Reference to the Region Within 3 Kiloparsecs of the Center," Proc. Nat. Acad. Sci. Wash. 46, pp. 1.
Oort, J.H. and Spitzer, L. 1955, "Acceleration of Interstellar Clouds by O-type Stars," Astrophys. J. 121, pp. 6.
Van der Kruit, P.C., Oort, J.H. and Mathewson, D.S. 1972, "The Radio Emission of NGC4258 and the Possible Origin of Spiral Structure," Astr. Astrophys. 21, pp. 169.

W.B. Burton, a former student of Peter van de Kamp at Swarthmore College, did his graduate studies at Leiden, 1963-1970, has been at the National Radio Astronomy Observatory and is now at the University of Minnesota.

W.B. Burton

ON HIGH-ENERGY ASTROPHYSICS

V.L. Ginzburg

Jan Oort is the same age as our century, the century of brilliant astronomic discoveries. Contemporaries, true, are apt to underestimate the achievements of the past and to attach particular importance to the events they witness. In fact, however, astronomy and physics began rapidly developing more than three hundred years ago and there is no ground to consider the XXth century to be distinguished in the rate of growth, striking discoveries etc. But this is another topic and here I may only restrict myself to the remark that a mere enumeration of the achievements connected with the name of Oort - galactic rotation, atomic hydrogen radioastronomy, study of the central region of the Galaxy, Crab Nebula, a number of galaxies, comets etc., - speaks for itself.

1. High-energy astrophysics has undoubtedly appeared only in our century and, in fact, not until its second half. It includes the establishment and account of the role played in astronomy by the cosmic rays and by the radio, optical, X-ray and gamma radiation and also the high-energy neutrinos produced by the cosmic rays.* It is quite obvious that a revolution in astronomy -its becoming all-wavelength rather then only optical astronomy- is closely connected with the development of high-energy astrophysics. This transformation and the discovery of the expansion of the Metagalaxy are most important events in the astronomy of the XXth century.

*"High-energy astrophysics" is often called "cosmic-ray astrophysics" or "the problem of the cosmic-ray origin". It is commonly accepted at present, however, that it is only charged particles that are referred to as cosmic rays. Therefore, the name "high-energy astrophysics" is better suited here, though as far as terminology and classification are concerned, it is difficult to achieve unambiguity and unanimity.

H. van Woerden, W. N. Brouw, and H. C. van de Hulst (eds.), Oort and the Universe, 129–140.

Below I shall dwell briefly on the origin and development of high-energy astrophysics. In most cases I shall not mention names and not refer the reader to the original literature since, in general, "priority questions are dirty business", and the character of the present book makes these questions particularly out of place*.

2. The study of cosmic rays began, in fact, somewhere near 1900, when gas ionization in closed vessels was observed: the question arose whether this ionization can be fully explained by radioactive radiation from the Earth's surface, from the vessel's walls and the radioactive emanations in the gas. It was not easy to solve this problem, especially not the role of the Earth. To this end some flights on balloons were undertaken, which led to an undoubted discovery of cosmic rays by V. Hess in 1912. Particularly successful was his flight on August 7, 1912, when he reached a height of 5 km; by that time the ionization rate had already increased several times as compared with that observed at sea level. These results were confirmed in 1914 by W. Kolhörster, who reached a height of 9 km, where the ionization rate is even greater. However, mentioning "undoubted discovery", we implicate the establishment of facts and to a certain extent the modern concepts of the structure of atmosphere. It was supposed, for example, that an increase of the ionization rate far from the Earth could be due to the presence in the upper atmosphere of a considerable amount of radioactive emanations rather than to the action of some unknown cosmic radiation. In any case it was not until about 1927 that all the doubts about the existence of cosmic rays, i.e. a penetrating "radiation" of extra-terrestrial origin, left because these "rays" proved to be absorbed much less than γ-rays from radioactive elements. But though the very existence of cosmic rays had been in doubt for at least 15 years, their origin was considered then to be clear - they "must be" hard γ-rays (such a conclusion was drawn because γ-radiation of radioactive elements is the most penetrating). But in 1927 a geomagnetic effect (the latitude-dependence of the ionization rate) was revealed and then studied. As a result, approximately by 1936 it had become clear that primary cosmic rays are charged particles. At first, electrons were taken for these particles, but later on protons were found to play the leading role. Finally, in 1948 the nuclei of a whole number of elements were detected in the composition of primary cosmic rays.

---

*A whole number of original papers, to say nothing of many references, can be found in Hillas (1972) and Rosen (1969). See also the reviews Ginzburg and Syrovatskii (1964) and Ginzburg (1978). An insight into the contemporary state of the problem is best of all obtained by getting acquainted with the Proceedings of the last Cosmic Ray-Conference, Kyoto (1979).

Thus, about 40 years passed before it was clarified in a very general outline what the cosmic rays are. I mention this, specifically, to stress that to solve some scientific problems takes us sometimes not less than several decades. Such a conclusion is, of course, not new. Nevertheless, even the events of 20 years, to say nothing of 30 years, ago often seem to be antiquities to the majority of scientists, because they are young.

Sometimes the events really move swiftly. For example, pulsars were very soon identified with neutron stars. It seems to me, however, that it happened only owing to the discovery of short-period pulsars in Crab and Vela. If only pulsars with a period larger then a second were known (like all the four first pulsars discovered in Cambridge), it would not be an easy task to choose between neutron stars and white dwarfs. Specification of the nature of central regions (kerns) of quasars may just serve as an example in this case. Quasars were discovered 4-5 years prior to pulsars, in 1963, but the nature of their kerns is not yet clear. I believe, though not everybody will support this, that the most probable kern model is a magnetoplasmic body (magnetoid, spinar) or a massive black hole. But the choice between these possibilities is so difficult that it may take some more decades for a solution.

3. As has already been mentioned, by 1950 the composition of primary cosmic rays had become known in outline. Some papers also appeared that foresaw the potential importance of cosmic rays for astrophysics. So, in 1934 Baade and Zwicky (1934) associated the appearance of supernovae with the formation of neutron stars and generation of cosmic rays. In 1949 Fermi approached the cosmic rays as a gas of relativistic particles moving in interstellar magnetic fields. Nevertheless, the role of cosmic rays in astronomy remained quite unclear on the whole, and, as far as I can judge, cosmic rays were of interest practically only to physicists. The main reason apparently lies here in a high degree of isotropy of the cosmic rays (disregarding the influence of the Earth's magnetic field). That is why even the most detailed information on the composition and energy spectrum of the cosmic rays near the Earth gives little evidence concerning the sources and particularly the localization of these sources. The situation is here analogous to that which would occur if only the spectrum of all the stars taken together were known, individual stars not being observed.

Therefore I think that the birth of cosmic-ray astrophysics and high-energy astrophysics as a whole happened in 1950-1953, when the situation changed radically. Namely, a considerable part of the cosmic radioemission turned out to be of synchrotron nature. As a result it became possible to obtain vast information about the electron component of the cosmic rays far from the Earth - within and beyond the Galaxy. Furthermore, making some assumptions one can estimate the total cosmic-ray energy in the sources (in the envelopes of supernovae, in radiogalaxies etc.) from the intensity of the synchrotron radiation. The comprehension of these points, speaking of the astronomic community as a whole, took about a decade but, in cany case, in the Paris Symposium on Radioastronomy in 1958 a close connection between radioastronomy and cosmic rays raised already no doubt. The history of this development* is rather dramatic and is often presented erroneously because of ignorance of the original literature (published sometimes only in Russian) and the use of versions "adopted by repetition" only. I shall restrict myself to the remark that a better insight into the role of the synchrotron mechanism in astronomy was to a considerable extent provided by the paper by Oort and Walraven (1956) devoted to the polarization of optical radiation from the Crab Nebula.

4. What basic conclusions were drawn, over 20 years ago, from the establishment of the connection between radio astronomy and the cosmic rays ?

First of all it became clear that the generation of cosmic rays is a universal phenomenon, since cosmic rays are present in the interstellar space, in the envelopes of supernovae and in other galaxies, particularly in radio galaxies. From this it follows also that cosmic rays are most valuable as a source of astronomical information, and not so much directly (observation of the cosmic rays near the Earth) but as a possibility to study radiation generated by the cosmic rays. At first the corresponding radio-emission alone was investigated. But then optical, X-ray, gamma-radiation and, in principle, high-energy neutrinos were added.

Secondly, the cosmic rays were found to be an important dynamic and energetic factor. Their energy density $W_{cr} \sim 10^{-12}$ erg cm^{-3} in the Galaxy is of the same order as the magnetic-field energy density $W_H = H^2/8\pi$ in the interstellar medium and the density of the internal (kinetic) gas energy $W_g = 1.5nkT$. In certain objects the density W_{cr} exceeds, or may exceed, the densities W_H and W_g. The same refers to the pressure of cosmic rays $P_{cr} = W_{cr}/3$ (in cosmic rays relativistic particles are predominant). The total energy of the cosmic rays in the Galaxy is estimated to be $W_{cr} \sim 10^{56}$ erg and in powerful radio galaxies $W_{cr} \lesssim 10^{61}$ erg $\sim 10^7 M_{\odot} c^2$.

Both the conclusions are in full accordance with modern plasma physics concepts: in the presence of particle beams, shock waves and various magnetic inhomogeneities, an effective acceleration of some part of the particles and also their scattering and diffusion should be expected in a rarefied plasma. The significance of the above-mentioned facts for astronomy can hardly be overestimated.

Since cosmic rays are such an important ingredient in space, the role of high-energy astrophysics as a whole is also quite obvious. Simultaneously, as happens usually in such situations, different scientific fields and trends penetrate into one another, and it is not an easy task today to establish the limits of high-energy astrophysics. Maybe it is not at all necessary.

---

* This story was briefly, but with the necessary references to the literature, presented in the introduction to the paper by Ginzburg and Syrovatskij (1965).

In the past decade the most important new trend in high-energy astrophysics seems to be the development and, as a matter of fact, even the very emergence of observational gamma-astronomy*.

Particularly significant from the point of view of the cosmic ray studies is the registration of γ-rays from Π^o-meson decay. These Π^o-mesons are produced in collisions of the protron-nuclear component of the cosmic rays with nuclei in a gas. The intensity $I_{\gamma\Pi^o}$ of these γ-rays is therefore proportional to the gas concentration n and to the intensity of cosmic rays I_{cr} or, after some recalculations, to their energy density W_{cr}. Thus, measurements of the quantity $I_{\gamma\Pi^o}$ offer, in effect, the only direct possibility to determine the density W_{cr} far from the Earth. It will be not out of place to draw here an analogy with the registration of synchrotron radiation, which makes it possible to establish the energy density of the electron component $W_{cr,e}$ (if the field H is known). One of the important results obtained by this gamma-astronomical technique will be touched upon below. We should also mention other possibilities of gamma-astronomy, for example, detection of nuclear γ-lines and annihilation radiation (the line $E_\gamma = 0.51$ MeV); registration of gamma-radiation due to the inverse Compton effect, for example in quasars; ground-based observations of γ-rays with $E_\gamma \gtrsim 10^{11}-10^{12}$ eV (by bursts of Cherenkov radiation in the atmosphere). In general, one can say that, account being taken of available experimental possibilities, the "gamma-window" to space is now widely open, and will undoubtedly be more and more used in astronomy.

5. One of the trends in high-energy astrophysics and, in fact, the oldest one, is the problem of the origin of cosmic rays. It has been discussed for already half a century (Rosen, 1969). Specially, we usually have in mind the origin of the bulk of the cosmic rays observed near the Earth. The origin of particles with a super-high energy ($E \gtrsim 10^{17}$ eV) and soft cosmic rays (sometimes referred to as subcosmic rays) with $E \lesssim 10^9$ eV are special questions.

To solve the problem of cosmic-ray origin means, first of all, to establish the "trapping region", wherein the energy density of the cosmic rays is of the order as in the vicinity of the solar system ($W_{cr} \sim 10^{-12}$ erg cm^{-3}). In metagalactic models, the trapping region is the entire Metagalaxy, the region like the Local Supercluster etc. In galactic models the trapping region (for the halo model) is a quasi-spherical or a somewhat flattened "cosmic-ray halo" with the characteristic size $R \sim 10$ kpc. For disk galactic models the trapping region is a disk with the half-thickness $h \ll R$. Another urgent question is the origin and localization of the cosmic-ray sources. Other numerous pro-

---

*Achievements of X-ray astronomy are even more significant, but this field is in a certain sense "not typical" of high-energy astrophysics. The point is that most of the cosmic X-ray emission is bremsstrahlung (radiation) of a hot but nonrelativistic plasma ($T \lesssim 10^9$ K $\sim 10^5$ eV) and is not directly connected with cosmic rays.

blems have also been discussed for a long time. The following ones may be mentioned here: the mechanisms of particle acceleration, the propagation of cosmic rays in interstellar space (diffusion and the conditions of applicability of the diffusion approximation, plasma effects, transformation of the chemical composition, losses, etc.), generation of secondary electrons and positrons etc. But the most important problem is to choose at least the type of the model, without which the question of the cosmic-ray origin remains open. This choice, however, required 25 to 30 years, even if we start from 1950-53 and disregard earlier attempts and models of the solar type (Rosen, 1969). Like a number of other physicists and astronomers I have been convinced of the validity of the halo galactic model from the very start (at least since 1952-1953). There was every reason for accepting this very model (for more detail see Ginzburg and Syrovatskij, 1964), but for a long time we had no direct and weighty proofs.

Now we have them.

After the discovery in 1965 of relict radioemission with a temperature $T_{ph} \approx 2.7$ K (the energy density $W_{ph} \approx 4 \times 10^{-13}$ erg cm^{-3}) it became clear that the electron component of the cosmic rays (at least for energies $E \gtrsim 10^{10}$ eV) must be of galactic origin; because of the inverse Compton losses on relict radioemission, the electrons cannot reach us even from the nearest radio-galaxies. This confirmed, though indirectly, the confidence in the galactic origin of the main, proton-nuclear component of the cosmic rays*. A direct proof may be obtained by the gamma-astronomical method. In metagalactic models the energy density W_{cr} outside the Galaxy is almost the same as within the Galaxy, i.e. of the order of $W_{cr} \sim 10^{-12}$ erg cm^{-3} (we mean at distances of $z \ll 1$). Therefore, one can predict, for example, the flux of γ-rays from the Π^0-meson decay produced in the Magellanic Clouds (this flux is determined by the value of W_{cr} and the amount of gas in the Clouds, which is known). The measurements of γ-ray intensity in the direction of the galactic anticentre and in neighbouring directions are somewhat less obvious but may prove to be quite convincing. Knowing the amount of gas along the corresponding lines of sight, one can indicate, say, the intensity $I_{\gamma,\Pi^0}(E_\gamma > 100\,\mathrm{MeV})$ Unfortunately, the Magellanic Clouds have not yet been observed in γ-rays, but the galactic value of I_γ was measured to be twice as large as the calculated value of I_γ for W_{cr}= constant. When we take into account the contribution from some disregarded γ-ray sources, for example, from discrete sources, this difference increases. The only possible explanation is a decrease of W_{cr} when moving away from the galactic centre, which contradicts the metagalactic model. Further observations and verifications are desirable, of course, but on the whole "the work is done".

---

*Electrons were observed in the primary cosmic rays only in 1961, the corresponding energy density $W_{cr,e}$ being $\sim 10^{-2}\, W_{cr} \sim 10^{-14}$ erg cm^{-3}. This result by no means contradicts radioastronomical data.

If there exists "the cosmic-ray halo", it is natural to expect the presence of a radio-halo, for which the electron component of the cosmic rays in the halo is reponsible. To reveal the radio-halo turned out to be very difficult, however, for which there is good reason. Nevertheless some radioastronomers seemed to be greatly irritated by this fact. The character of the debates that took place in 1966 is clear, for example, from van Woerden (1967). I believe that in recent years the question has been quite convincingly clarified as a result of the discovery of a radio-halo in on-edge galaxies NGC 4631 and NGC 891 (Ekers and Sancisi, 1977; Allen, Baldwin and Sancisi, 1978). By the way, these measurements have been taken by the unique radiotelescope in Westerbork which was created, as I have heard, only due to the efforts of Jan Oort. For the Galaxy the results are not so obvious, but treating the radio-data by different methods one also comes to a positive conclusion about the existence of the radio-halo (Bulanov, Dogel and Syrovatskii, 1976; Webster, 1978). Note, by the way, that the arguments in favour of the disk galactic model (with a rather thin disk, in which the lifetime of cosmic rays is less than 10^7 years) presented in the literature are, in fact, the result of a misunderstanding (Ginzburg, 1978).

Referring the reader for some details and additional literature to Ginzburg (1978) and Kyoto (1979), we draw a general conclusion: the model of the origin of the bulk of cosmic rays observed near the Earth is a halo galactic model. In such a model the characteristic energy, lifetime and power of generation of the cosmic rays make up $W_{cr} \sim 10^{56}$ erg, $T_{cr} \sim 1\text{-}3 \times 10^8$ years, $U_{cr} \sim W_{cr}/T_{cr} \sim 1\text{-}3 \times 10^{40}$ erg/sec.

As to the cosmic-ray sources, the candidature of supernovae, put forward in 1934 (Baade and Zwicky), is still a favourite. Radio-observations of supernova remnants for almost 30 years leave no doubt as to the presence of relativistic electrons in these remnants. But the role of the proton-nuclear component of the cosmic rays in the remnants is still unclear. Progress in this direction may be expected first of all from the use of gamma-astronomy.

Strictly speaking, there is no doubt that directly or indirectly (we mean particle injection and also acceleration in the interstellar space by shock waves coming from supernovae) supernovae generate a considerable part of the galactic cosmic rays. But some contribution can be made by other stars as well. In recent years O-stars are particularly popular in this respect (Kyoto, 1979). What is this contribution? There are no convincing data, and the answer, evidently, depends on the energy. My opinion (on the basis of the arguments presented, for example, already in Ginzburg and Syrovatskij (1964)) is as follows: at energies $E > 10^{10}$ eV supernovae dominate as the sources, although this has not been proved. The contribution from various other stars is in any case of interest, and if it would prove to be significant, it would even enrich the picture. It should be noted that apart from the above-mentioned gamma-method, the nature of the sources, of course, has an effect on the chemical and isotopic composition of the cosmic rays. The appropriate measurements of the composition are being improved and will make their contribution to the solution of the problem of sources.

6. It seems to me that cosmic-ray astrophysics, or, if you like, high-energy astrophysics has just come or is just coming (there is no sharp boundary) to a turning point.

The cosmic-ray origin model (in a restricted formulation - for the bulk of the cosmic rays observed on the Earth) is clear in its general outline. Gamma astronomy stands on its own feet, has already yielded real results and promises still more achievements.

In any case, old "damned questions" are replaced by new ones which are not easy to answer. What are these questions ? How will high-energy astrophysics develop and what will it yield, say, by 2000 ? We are, of course, very much restricted in our ability to glance into the future. But 10 or even 20 years are not such a long period for the development of science, as I tried, by the way, to illustrate above. At the present time even an additional obstacle has appeared which is connected with the complexity of the equipment. Designing and putting such equipment into operation (for example, on a satellite) are usually separated by a decade.

There is no doubt that within the coming 10 to 20 years we are going to witness some unexpected events and probably even great discoveries. This unexpectedness is one of the charms of science. At the same time, it is also useful to try to foresee some things, and I would like to express very briefly my opinion concerning some perspectives of high-energy astrophysics.

a. The study of the chemical and isotopic composition of the cosmic rays is now at a turning point. A new generation of modern equipment for satellites and high-altitude balloons has already been born or will soon appear. Hence, in this field one can expect great and rather rapid progress. I mean the establishment of the composition and energy spectrum of nuclei up to the energies of 10^{12} -10^{13}eV nucleon^{-1}. The isotopic composition will be known only at lower energies, but even, for example, finding the amount of radioactive nuclei of ^{10}Be at $E \gtrsim 10Mc^2$ will be a great achievement (the mean lifetime of the ^{10}Be nucleus is $\tau = 2.2 \times 10^6$ years; such nuclei play the role of a clock, and the measurement of their amount provides information about the age of the cosmic rays (Ginzburg, 1978; Kyoto, 1979)).

Sufficiently detailed data on the chemical and isotopic composition of cosmic rays will make it possible, although there are considerable difficulties, to find out the composition of cosmic rays in the sources. This is undoubtedly one of the important sources of information.

b. The spectrum of cosmic rays is known up to energies of about 10^{20} eV. At super-high energies ($E > 10^{16}$-10^{17} eV) there are so few particles* that they can be observed only by extensive showers in the

---

* The intensity of primary particles with an energy $E \geq 10^{16}$ eV makes up about 10^2 km^{-2} sterad^{-1} h^{-1}. At $E \gtrsim 10^{20}$ eV this intensity is already equal to about 10^{-6}km^{-2} sterad^{-1} h^{-1} ~10^{-2}km^{-2} sterad^{-2} y^{-2}.

atmosphere. The chemical composition of cosmic rays of super-high energy is known insufficiently, their origin is also unclear. Neither galactic nor extragalactic origin is excluded at present (by extragalactic we imply the local Supercluster; particles with the observed energy spectrum cannot come from more remote regions because of the losses due to relict radiation (Berezinsky, 1977)). The model, in which particles with $E<10^{19}$ eV are mainly galactic and at $E>10^{19}$ eV come already from the local Supercluster (Kyoto, 1979; Berezinsky, 1977), seems most probable at the present time. One may hope that the problem will be solved (in the sense of tentative localization, origin, etc.) within 10 to 20 years (one of the most important ways is the measurement of the anisotropy of the cosmic rays with super-high energy).

Apart from its significance for astrophysics, the study of cosmic rays with super-high energy is now, and in all probability will for a long time remain, important for physics. Remember that from 1927-29 up to the early fifties the cosmic rays were widely used in high energy physics, and they helped to discover (Rosen, 1969) the positron e^+ (1932), $\mu^{\pm}$-mesons (1937), $\Pi^{\pm}$-mesons (1947), K^0 and $K^{\pm}$ -mesons (1947-48) and Λ, Σ^+ and Ξ^--hyperons (1951-53). But since then the centre of gravity of the corresponding physical researches has shifted towards accelerators. This is quite comprehensible, and if for a given energy E one can utilize an accelerator, then cosmic rays cannot compete . However, in the eighties one can evidently rely, as a maximum, upon the use of colliding proton beams with an energy $E_c=10^{12}$ eV per beam (project of the Fermi Laboratory, USA). In the laboratory system it is equivalent to the proton energy $E = 2E_c^2/Mc^2 \simeq 2x10^{15}$ eV. So, at energies $E > 2x10^{15}$ eV the cosmic rays are the only source of particles and this may be valid till the end of the century. Of course, to carry out physical researches with cosmic rays in the energy range $E > 2x10^{15}$ eV is very difficult, but research has always encountered difficulties, and scientists are now equipped with equipment that our predecessors had never even dreamt of.

c. A number of remarks concerning gamma-astronomy have already been made above. This scientific branch may also be considered to be passing through a turning-point. The current decade will probably lead to some progress analogous to that of the seventies made in X-ray astronomy (the culmination in this field is considered to be the results obtained by the "Einstein" Observatory (Astrophys.J. Letters, 1979)). A new generation of gamma-telescopes will make it possible not only to refine the results obtained from the satellites SAS-2, COS-B etc., but also to explore a great number of discrete sources, including the Magellanic Clouds, a number of galaxies and their nuclei and quasars. Sufficiently impressive are even the results already available (Kyoto, 1979). For example, the statement about γ-luminosity of the quasar 3C 273 estimated as $L_\gamma(50<E_\gamma<500$ MeV$) = 2x10^{46}$ erg s^{-1} (the distance R = 790 Mpc is assumed). For 10^6 years such a luminosity corresponds to the energy $W_\gamma \sim 6x10^{59}$ erg $\sim 3x10^5$ $M_\odot c^2$ radiated in γ-photons alone. The optical and X-ray luminosity of this quasar is approximately the same as its gamma-luminosity and only in the infra-red region is the luminosity higher by an order of magnitude. For the pulsar PSR 0532 (the Crab pulsar) $L_\gamma(E_\gamma>100$ MeV$) \simeq 3.5x10^{34}$ erg s^{-1}. For the source

Cyg X-3 (this may be a young pulsar in a binary system) $L_\gamma(E_\gamma>40$ MeV) $\simeq 7.5\times10^{36}$ erg s^{-1}, $L_\gamma(E_\gamma>10^{12}$ eV) $\simeq 10^{35}$ erg s^{-1}.

Large gamma-luminosities are rather significant and, in any case, testify to a large amount of cosmic rays (electrons, in the case of the Compton mechanism, and also synchrotron and curvature radiation in the case of pulsars). Other branches of gamma-astronomy (radiation in lines, etc.) may also yield many interesting results.

High-energy astrophysics is closely connected with investigations in X-rays and, sometimes, in other wavelength bands. But it is quite clear that the advancement is on a wide front, and here I would like to underline only some key points.

d. Such key points should include high-energy neutrino astronomy. This field, if we mean experiment, is only emerging. However, underground measurements of neutrinos, say, from supernova flares in our Galaxy are already quite feasible. The creation of deep under-water optical and (or) acoustic systems (project DUMAND, etc.) will make it possible to fix confidently and with a rather high angular resolution (of the order of 1°) neutrinos with energies $E_\nu>10^{12}$ eV from remote extragalactic sources (for some estimates and literature: Kyoto, 1979; Berezinsky and Ginzburg, in press). Neutrinos of the above mentioned energy are created in practice by the proton-nuclear component of cosmic rays only and, consequently, may serve as its indicator (analogous to γ-rays from Π^{o}-meson decay in a less hard spectral region). Besides, neutrinos possess a large penetrating ability. At the same time even γ-rays with energies $E_\gamma \gtrsim 2\times10^{11}-10^{14}$ eV are already strongly absorbed at large metagalactic distances (the process $\gamma+\gamma^{1}\rightarrow e^{+}+e^{-}$, where the role of γ^{1} is played by soft photons of relict and optical radiation). Gamma-rays are even more readily absorbed by a layer of matter with a thickness of 100 g cm^{-2}. Therefore, they cannot escape from the internal regions of, say, dense galactic nuclei. This, precisely, explains to a considerable extent the difficulties encountered in establishing the nature of the kerns of quasars and active galactic nuclei. Observation of the high-energy neutrino radiation from these objects along with gamma-astronomical observations offers for some models the possibility of distinguishing between a massive black hole and a magnetoid (Berezinsky and Ginzburg, in press).

In general, I am sure, high-energy neutrino astronomy* is the principal, as yet, unused reserve of high-energy astrophysics and astronomy as a whole (though in the latter case I cannot but mention another not less important "reserve": gravitational-wave astronomy).

Modern astronomy already seems inconceivable without high-energy astrophysics. By the end of the century it will be so obvious and commonly accepted that it will not even need special mentioning (possibly this is not quite necessary even today).

---

* Low-energy neutrino astronomy ($E_\mu \lesssim 10$-20 MeV) is, of course, also one of the most important areas in the study of the Sun, supernova flares and, possibly, some other objects.

7. In conclusion, may I permit myself to make some remarks of a personal nature.

I saw Jan Oort for the first time in the middle of 1947 in Leiden, but I did not even make his acquaintence then ! It was a coincidence that a group of Soviet astronomers and physicists came to Leiden - on the way back from an expedition to Brazil organized for observing by radio and optical methods the total solar eclipse on May 20, 1947. My participation in the expedition was, in fact, a prize for the work on radiowave propagation in the ionosphere and solar radioemission. It is to the latter topic that my first and, at that time, single astronomical paper (Ginzburg, 1946) was devoted. Astronomy of a wider scope was then unknown to me, and if I am not mistaken, I knew nothing or next to nothing about Jan Oort and his achievements and activity. I was much more attracted by low-temperature physics and that is why, when our astronomers started for the Observatory, I rushed to the well-known Kamerlingh Onnes cryogenic laboratory. These details are, of course, not very interesting, but I mention them because up to now I feel regret for the missed opportunity to talk with J. Oort. It seems possible to me that if that had happened, I would not have waited for some more years before coming close to astronomy again in 1950-1951.

Since then I have turned, and for some time even at the expense of my work in physics, precisely to high-energy astrophysics. From that time the name of Oort means much to me. I became acquainted with him personally during one of his visits to the USSR. And then we met at the IAU Symposium in Noordwijk in 1966. It was a very remarkable symposium, at any rate I remember it as such. It is difficult to say, however, whether this was due mostly to science, or to the beauty of Holland, the perfect organization, friendliness and kindness of the hosts headed by J. Oort. At the same time it is clear from the proceedings of the Symposium (van Woerden, 1967) how interesting it also was in a scientific respect. There were some more meetings later on. In 1975 J. Oort, who was elected a foreign member of the USSR Academy of Sciences in 1966, came to Moscow together with his wife to celebrate the 250-th anniversary of the Academy. In 1976, during my last trip to Europe I also saw J. Oort but since then we have not met. However, the stay in Leiden, where I was with my wife for a whole month, I remember particularly well. We lived in the old building of the Observatory right above the Oort's flat. We often met the Oorts and our feelings towards them became and will always remain very warm.

The question arises again, what are all these details for? I myself am not sure that they are not superfluous. But still I wish to mention that there are reasons why I write differently from my Western colleagues who do more travelling. I will not have the opportunity to congratulate the hero of the day personally on April 28, 1980. But at the same time, it is even easier to express the deep respect and love I feel towards Jan Oort in writing.

References

Allen, R.J., Baldwin, J.E. and Sancisi, R. 1978, Astron. Astrophys. 62, pp. 397,
Astrophys. J. Letters 1979, Volume 234, part 2.
Baade, W. and Zwicky, F. 1934, Phys. Rev. 46, pp. 76.
Berezinsky, V.S. 1977, 15th International Cosmic Ray Conference, Conference Papers 10, pp. 84.
Berezinsky, V.S. and Ginzburg, V.L. 1980, Mon. Not. Roy. Astron. Soc. (in press).
Bulanov, S.V., Dogel, V.A. and Syrovatskij, S.I. 1976, Astrophys. and Space Sci. 44, pp. 267.
Ekers, R.D. and Sancisi, R. 1977, Astron. Astrophys. 54, pp. 973.
Fermi, E. 1949, Phys. Rev. 75, pp. 1169.
Ginzburg, V.L. 1946, Doklady Akad. Nauk USSR 52, pp. 487.
Ginzburg, V.L. 1978, Sov. Phys. Uspekhi 21, pp. 155.
Ginzburg, V.L. and Syrovatskij, S.I. 1964, The origin of Cosmic Rays (Pergamon Press).
Ginzburg, V.L. and Syrovatskij, S.I. 1965, Ann. Rev. Astron. Astrophys. 3, pp. 297.
Hillas, A.M. 1972, Cosmic Rays (Pergamon Press).
Kyoto 1979, 16th International Cosmic Ray Conference, Conference papers
Oort, J.H. and Walraven, Th. 1956, Bull. Astron. Inst. Netherlands 12, pp. 285.
Paris Symposium on Radio Astronomy 1959, Proceedings (Stanford University Press).
Rosen, S. (ed.) 1969, Selected papers on Cosmic Ray origin theories (Dover Publication).
Van Woerden, H. (ed.) 1967, Radio Astronomy and the Galactic System (IAU Symposium 31, Academic Press).
Webster, A. 1978, Mon. Not. Roy. Astron. Soc. 185, pp. 507.

V.L. Ginzburg is at the Lebedev Physical Institute and has been guest professor at Leiden in 1969.

V. Ginzburg

OORT AND EXTRAGALACTIC ASTRONOMY

Margaret and Geoffrey Burbidge

1. Introduction

Our understanding of the nature of external galaxies, incomplete as it still is, has depended most essentially on the understanding of our own Milky Way. And one cannot think of any facet of knowledge about our Galaxy without paying tribute to Jan Oort's elucidation of its mysteries. One especially thinks of his remarkable - yet how logical! - discovery that the Galaxy is a flattened assemblage of stars and other matter in differential rotation under the influence of its own gravitational field.

This is easy to comprehend now, but in looking back over more than half a century of fundamental research by Oort, one can have an inkling of the excitement he must have experienced when it became clear to him that the motions of stars in the solar neighborhood and in the middle distance from the Sun could be interpreted as rotation of the whole system. There followed his beautiful mathematical analysis of these motions, and their kinematical interpretation as a three-dimensional velocity field with the major component due to rotation of the system (Oort 1927, 1928). In addition to the dynamical analysis of the rotation, Oort recognized the very important component of velocity perpendicular to the Galactic plane as due to harmonic motions resulting from the force field of the matter in the Galaxy, from which the local mass density could be derived (Oort 1932, 1960).

These discoveries are part of the structure upon which subsequent work on external galaxies has been built. The Oort A and B constants, the determination of the distance of the Sun from the Galactic Centre, and the determination of the local density of matter and of the mass of the Galaxy as a whole, are subjects for other papers in this Festschrift. But we wish to emphasize the importance of these concepts for all subsequent work on the dynamics of galaxies.

It is not easy to give a full account of Oort's work in extragalactic astronomy, because it has been so wide-ranging and has encompassed the entire half-century of work on the physical nature of galaxies, their origin and evolution, and the structure of the Universe as depicted by them. Topics we wish to cover (however inadequately) are the follo-

H. van Woerden, W. N. Brouw, and H. C. van de Hulst (eds.), Oort and the Universe, 141–150.

wing, although not precisely in the order listed here. First, the masses of galaxies, their mass-to-light ratios, and the mass-to-light ratios of clusters of galaxies. This is linked, of course, to the fundamentally important work on the 21-cm H I line, but it is also connected with the whole subject of radio galaxies, because the same basic equipment is used for both. Spanning both of these main categories of subjects is that of the systematics of spiral galaxies, and the problem of galactic evolution. And interconnected with all of these is the very important subject of the nuclei of galaxies: the nuclear structure and its relationship to the rest of the galaxy, the energy source for activity in the nucleus, the interconnection between nuclei and the giant extended radio sources of radio galaxies, and the generation of non-circular motions in and near the nuclei of active and even relatively quiescent galaxies. The latter topic connects with questions of the origin of all unusual motions in galaxies, from the high-velocity H I clouds in our own Galaxy to those in very distant galaxies, and is related to the question of the interaction between galaxies and whatever ambient medium surrounds them.

The following sections will attempt to divide up these interconnected studies in the best way we can, while keeping always to the forefront the ever-present connecting chain: Jan Oort's goal of understanding the Universe.

2. The 21-cm hydrogen line in galaxies near and far

We wish to discuss this topic ahead of the subject of masses of galaxies, because of the immense importance of the 21-cm line in determining masses. Thus, we are moving ahead in chronology to 1944, in the dark days of World War II in occupied Holland, where resistance to the Nazis, and aid or support to the resistance movement, was punishable by death and worse.

The prediction in 1944 that a line of radio wavelength should be emitted or absorbed by cold neutral hydrogen was made by Van de Hulst. The publication of this prediction in 1945 of course paved the way for a worldwide effort to observe the line. Not surprisingly, the U.S. was able more quickly to muster the equipment than was war-torn Holland. Thus, the first observation was made by Purcell and Ewen on 25 March 1951, but it was followed rapidly by the observation made on 11 May 1951 at Kootwijk (Muller and Oort, 1951). And it is noteworthy that, while Purcell and Ewen essentially dismissed the relevance of their observation to the possibility of measuring rotational velocities in the entire Galaxy, because of the "absence of large frequency-shifts" in their observations, Muller and Oort immediately recognized this important possibility. Muller and Oort indeed measured frequency shifts, and compared the magnitude of the velocities obtained with those given by a mass model of the Galaxy (Oort 1941). In the latter paper, interestingly enough, Oort had commented on the mass distribution in NGC 3115 in comparison with that in our Galaxy, on the basis of work which we shall describe in Section 3.

After these initial observations, work leapt ahead in both the northern and southern hemispheres; spiral arms in the Galaxy were delineated (e.g., Oort, Kerr and Westerhout (1958), which is a "Report on the Progress of Astronomy" on "The Galactic System as a Spiral Nebula"), and it was not long before the technique was applied to external galaxies. Van de Hulst, Raimond and Van Woerden (1957) published detailed observations of M31 made at Dwingeloo, and the Dutch group has remained in the forefront of extragalactic observations of 21-cm radiation, first with the Dwingeloo radio telescope, and more recently with the 1.5-km synthesis telescope at Westerbork. It is obvious that these two facilities would not have come into existence without the drive and inspiration of Jan Oort.

While there are review articles on observations of the 21-cm emission line in external galaxies (for both distribution of H I and rotation curves), and these reviews give full bibliographies (we note the article by Van der Kruit and Allen (1978) and the chapter by Roberts (1975) as reference sources), we wish to point particularly to the review paper by Oort (1974) which brought together results obtained up to that date by Leiden and Groningen astronomers at Westerbork. This paper is entitled "Recent Radio Studies of Bright Galaxies"; it discusses both the spiral structure and rotation curves of galaxies determined from 21-cm observations, and the continuum radio emission in these galaxies and its distribution. The galaxies discussed in detail by Oort (1974) are M51, M81, and NGC 4258 (we shall return to a discussion of NGC 4258 in Section 5). He pointed out especially the greatly increased H I density in the spiral arms, as compared with the interarm regions, and the decrease in H I density along the arms as these are followed toward the centre. He went on to discuss the implications of these observations for the study of galactic evolution, since the regions of highest gas density will be the regions where star formation is currently taking place.

All the observations described so far in this chapter refer to the 21-cm line in emission. However, it can equally be detected in absorption if cold H I gas lies in front of a radio source. This is increasingly proving to be an excellent way of probing the very diffuse gas around galaxies, by using background radio sources of high brightness temperature which are located in the fields of the galaxies to be studied. Also, for those galaxies with strong continuum radio sources of small angular diameter in their nuclei, the 21-cm line in absorption can be an important clue to understanding what is going on in that galaxy, as we shall see later in NGC 1275 (Section 5). Most recently, the detection of 21-cm absorption in a few QSOs has been achieved, and a new field is opening up through the increased capability of today's radio telescopes in detecting weak 21-cm absorption at large redshifts.

In this section our intention has been to describe the way in which the discovery of the 21-cm line of H I has revolutionized extragalactic astronomy. It is fair to point out, despite the large number of astronomers now engaged in these observations, that if any one person can be identified as the leader and goal-setter in these studies, it is Jan Oort.

3. Masses of galaxies, mass-to-light ratios, and the mass density in the Universe

This is a very big subject, containing as it does so much of Oort's work in extragalactic astronomy, and our account will not, we fear, do full justice to his contributions. In Section 1 we noted the debt that all studies of masses of individual rotating galaxies owe to Oort's analysis of our own Galaxy. In this section we will try to pick some highlights in the steadily increasing body of knowledge on masses of galaxies, and the application of these data to the determination of mass-to-light ratios in galaxies and of the mass density in the Universe.

First, let us identify what we see as the main epochs into which the whole body of knowledge can be divided. The earliest indications of rotations of galaxies, from tilted spectrum lines, date back to the observations of Slipher, Wolf, and Pease, which of course predate the acceptance that the "nebulae" were star systems like our own. There was then a hiatus until the 1930's, by which time Oort's analysis of the rotation of our Galaxy had been well digested and improvements in photographic plates, slow though these still were, had made it feasible to attempt observations on external galaxies. References to studies in this period may be found in Schwarzschild (1954), Burbidge and Burbidge (1975), and Van der Kruit and Allen (1978).

The paper in this period to which we wish to draw attention is that by Oort (1940) entitled "Some Problems Concerning the Structure and Dynamics of the Galactic System and the Elliptical Nebulae NGC 3115 and 4494". This paper had been presented by Oort at the dedication of the McDonald Observatory. What Oort did was to combine surface photometry from photographic plates taken by Oosterhoff at Mount Wilson with Humason's unpublished (and apparently never to be published) measurements of rotational velocities from absorption lines in NGC 3115. These measurements extended only to 45 arc sec from the centre; even so, presenting a linear velocity curve as they apparently did over a region in which the surface luminosity was falling, they were perceived by Oort as posing a fascinating problem for interpretation. The linear curve indicated an approximately constant mass density; this, combined with Oort's measurements of luminosity in NGC 3115 which showed a normal elliptical-type decline, led him to conclude that *the distribution of mass appears to bear almost no relation to that of light*. The mass-to-light ratio in solar units, which he had earlier recognized to be a very important indicator of the type of stars emitting the light, must, he concluded from Humason's data, rise in the outer parts to a large value. The mass of NGC 3115, he deduced, must be at least $5 \times 10^{10} M_{\odot}$, possibly larger. While the surface photometry of NGC 3115 was done again photoelectrically at McDonald Observatory (Van Houten, Oort, and Hiltner 1954), along with several other nearby galaxies, the puzzle remained, although Humason's early observations were later improved.

This epochal paper (Oort 1940) appeared in a volume of the Astrophysical Journal in which there were only four other papers on objects outside our Galaxy! (The four were two by Seyfert, one by Carlson on some corrections to the NGC and IC, and a paper by Milne on cosmological theories.) What a comparison this makes with today's Astrophysical

Journal, so heavily weighted toward extragalactic research!

Oort's 1940 paper was used by Schwarzschild (1954), who analyzed the then sparse information available on masses and mass-to-light ratios of galaxies, and used Oort's methods of analysis on a newer (and again unpublished!) rotation curve of NGC 3115 by Minkowski. Incidentally, it was Schwarzschild's 1954 paper which set the two of us on the path of observing galaxies for the determination of velocity curves, mass distributions, and masses; we were able to start this work in 1958 at the McDonald Observatory.

Oort returned to the subject of NGC 3115 in his George Darwin Lecture (Oort 1946), where he rejected the possibility that the high mass-to-light ratio might be due to interstellar dust, and concluded that the outer regions of NGC 3115 must be *made up of extremely faint dwarfs, which compared to the solar-type stars should be relatively at least 10,000 times as numerous as in the galactic system.* After a discussion of his surface photometry of the Sombrero galaxy, NGC 4594, he said: *the experience gained by the study of the elliptical nebula NGC 3115 warns us not to put much trust in calculations of the stellar density from the light.* How interesting this is, in the light of the recent observations of rotation curves of spiral galaxies that extend nearly flat beyond the Holmberg radius of the visible galaxy! But this is jumping ahead too far; we want to discuss the next era of mass determinations, from both optical and 21-cm observations, summarized by Burbidge and Burbidge (1975) and Roberts (1975).

It was during our own work on velocities in galaxies that we had our closest connection with the *Sterrewacht te Leiden* in its beautiful setting of canals and botanical gardens. Kevin Prendergast and ourselves used to spend part of our summers in Leiden, and it was there that we learned our true appreciation of Jan Oort as a scientific leader, as well as gaining the continuing reward of friendship with Jan and Mieke Oort. We used to bring our material acquired during the year, to work on with the benefit of extensive discussions with Jan Oort and the radio astronomers at Leiden.

The most recent events on the study of the measurement of rotations of galaxies have been due to the radio astronomers, using the 21-cm line, and to Vera Rubin and colleagues, using the Kitt Peak and Cerro Tololo 4-m telescopes. The earlier rotation curves often (although not in all cases) reached a flat maximum and began to decline, as one would expect, with Keplerian motions and a fairly constant mass-to-light ratio through these mainly Sc galaxies. However, observations of much fainter ionized hydrogen further out, and of H I well beyond the luminous extent of some galaxies, showed that the velocity curves could not be extrapolated downwards as had been assumed, but run off to large distances from the nucleus at nearly constant velocity. Unless non-Keplerian motions exist here, this indicates "unseen" mass, and returns us to the point made so many decades earlier by Oort: that one should not put much trust in calculations of the density of matter from the visible light!

While these observations do indicate larger masses for spiral galaxies than previously calculated, their main impact is the problem posed by the question: what is the nature of the unseen mass that appears to be present?

This leads to the final topic of this section: the density of matter in the Universe. In a well-known and very often referenced paper, Oort (1958) combined counts of galaxies of given magnitudes with average masses and estimated that the overall density of luminous matter in the Universe is 3×10^{-31} g cm^{-3} (for a Hubble constant of 75 km s^{-1} Mpc^{-1}), or about two orders of magnitude less than that needed to produce a closed Universe. This raised the famous question of the "missing mass", already indicated by work on the velocity dispersion of galaxies in the Coma Cluster, which led to a large total mass and a very large mass-to-light ratio for the cluster as a whole. We note that the existence of "missing mass" in our Galaxy was first indicated by Oort's comparison between the mass density in the solar neighborhood, calculated by his analysis of the z-motions of stars, with actual counts within the observed stellar luminosity function.

The cosmological questions raised by the study of the density of matter in the Universe are still unsolved. This brief account leads fairly naturally to the problem of the origin of galaxies, the subject of a detailed paper by Oort (1970). However, we want to discuss this in the next section.

4. Systematics of spiral galaxies; formation and evolution of galaxies

Rotation curves of spiral galaxies available by 1961 were amply sufficient to show that these galaxies are in differential rotation. How, then, could spiral structure persist during some 100 revolutions of a galaxy during its lifetime? This problem of the persistence of spiral structure was addressed by Oort (1962). He showed the magnitude of the problem, and considered some possible solutions. These included systematic flows inward or outward, such that in a suitably chosen frame of reference the spiral arms would be fixed structures, or the possibility that matter might be constantly added to the inner edges of spiral arms, while the outer edges would constantly lose matter. C.C. Lin has acknowledged that this paper was the starting point for his density-wave theory for the maintenance of spiral arms.

Turning to the spiral galaxy closest to hand, our own, we look at a puzzling phenomenon discovered at Dwingeloo and analyzed by Oort in a series of papers (Oort 1966, 1967, 1969): that is, the existence of high-latitude, high-velocity clouds of H I that are approaching with velocities clearly not due to galactic rotation. These might be actual infall of intergalactic gas, or the return of gas ejected by supernovae, or some kind of circulation through a gaseous halo. In a detailed paper on the formation of galaxies and the origin of this gas, in which he produced a "grand scheme" for the evolution of galaxies, Oort (1970) reviewed the observations and concluded that inflow is occurring at a rate that would increase the mass of the Galaxy by 0.9% per 10^9 years.

In the 1970 paper, Oort put forward the hypothesis that galaxies formed in a Universe that had a high degree of turbulence, as earlier proposed by von Weizsäcker, and obtained their angular momentum from this initial turbulence. Starting with a simple schematic model, he derived equations giving the evolution of a protogalaxy from a turbulent element in an early Universe to a galaxy as presently observed. In this important and comprehensive paper he integrated the observed parameters of galaxies - morphological forms, masses, rotations - with conditions in an early Universe, also the properties of clusters of galaxies and the observed frequency of radio galaxies.

Results from this paper were combined with those to be discussed in the next section into a "Phenomenology of Spiral Galaxies" (Oort 1975), outlining the "grand scheme" and combining it with the distribution of the different types of stellar populations and the kinematics and dynamics of matter in the nuclear regions of galaxies.

5. Nuclei of galaxies; non-circular motions in galaxies; active and radio galaxies

This section starts with a relatively "inactive" galaxy, our own. The 21-cm observations made at Dwingeloo revealed surprising motions in the gas in the inner parts of our Galaxy. There is the now famous 3-kpc arm, moving away from the centre at 50 km s^{-1}, and an almost equally massive arm expanding away on the opposite side of the nucleus at 135 km s^{-1}; also a fast-rotating inner nuclear disk (Van Woerden, Rougoor, and Oort 1957; Rougoor and Oort 1960). There is also evidence for the ejection of two cones of gas in opposite directions from the nucleus, at an angle of about 45° to the plane of the Galaxy (Van der Kruit 1970); this was also discovered at Dwingeloo. A fine review of the complexities of the Galactic Centre, and also a less technical account, were published by Oort (1977a, b). These observations of our galactic nucleus, initiated under Oort's guidance, clearly demonstrate that even a fairly quiescent galaxy can have considerable activity in its nucleus. We refer also to Oort's discussion of the nucleus of M31 and an earlier account of the nucleus of our Galaxy, in which a comparison of the two systems is made (Oort 1971).

Activity on a more violent scale can produce large non-circular motions further out and might provide the initiating impulse for the formation of spiral structure. This was discussed by Van der Kruit, Oort, and Mathewson (1972) in a paper detailing the remarkable radio structure of the Sb galaxy NGC 4258. Non-circular motions and curious filamentary Hα arms had earlier been detected optically; the paper quoted above combined all these observations into a model in which clouds of gas amounting to 10^7 - 10^8 $M_{\odot}$ were ejected from the nucleus some 2×10^7 years ago at velocities of order 10^3 km s^{-1}. The radio observations were made with the Westerbork Synthesis Radio Telescope, and yielded detailed radio maps that correlated remarkably with the unusual optical structure. The hypothesis that events of this type and magnitude may initiate the formation of spiral structure is, of course, linked to the topic of Section 4.

On a smaller distance scale, i.e., closer to the nucleus, our own observations of unusual gas flow in the spiral galaxies M51 and NGC 253 revealed motions of order 100 km s^{-1} that were clearly non-circular and were probably located in cone-like structures inclined at quite large angles to the equatorial plane. We were sustained in our belief that this kind of model could represent these motions by the work of the Leiden astronomers on the Galactic Centre, and by those wonderful summer-time discussions with Jan Oort to which we have already referred.

The last case of "unusual" velocities in an active galaxy which we want to discuss is the Seyfert galaxy and strong radio source, NGC 1275. This galaxy, which has long been known to show gas at two velocities differing by 3000 km s^{-1}, is clearly a very active galaxy and its nucleus shows remarkable structure on a milli-arcsecond scale, from VLBI measurements. While we developed a model in which the gas at high positive velocity had been ejected from the active nucleus, Oort (1976) and Rubin, Ford, Peterson, and Oort (1977) adopted a model in which it was in fact a separate galaxy in the Perseus Cluster with large random motion, seen along the line of sight to NGC 1275. They drew this conclusion because of obscuring matter amongst the high-velocity gas which clearly lay _in front of_ NGC 1275; also 21-cm _absorption_ (i.e., cold H I) at the higher velocity had earlier been observed right in the nuclear radio source.

While we personally still think the high-velocity gas must be somehow connected with NGC 1275, we found these observations very interesting, and there is no denying that NGC 1275 is one of the most remarkable of active galaxies!

6. Conclusion

There is no conclusion. Always in the forefront of extragalactic research, Jan Oort has just promised a manuscript on the new problems to which he has turned: quasars and their absorption-line spectra. As we said at the beginning of this chapter, it is impossible to do justice to the massive contributions to knowledge of the Universe for which Jan Oort has been responsible, and the immense inspiration he has provided to all workers in the field.

We would like to sum up by pointing out that Jan Oort's influence on extragalactic research extends far beyond the work in which he himself participated actively. It has spread through the world as *a feeling for the facts* - a quiet, but rigorous, questioning that all of us who have come under his influence revere: What do you need to find out? What questions are you posing? How best can you obtain observations that will give answers? What equipment will you need to do this? How accurate will your answers be? How will you analyze them? What conclusions will you be able to draw from your analysis?

We are constrained to use a cliché to express how we see Oort's work in extragalactic astronomy: *The truth, the whole truth, and nothing but the truth* - enlightened by inspiration that comes only from an incisive and deeply reasoning intelligence.

References

Burbidge, E.M. and Burbidge, G.R. 1975, Chap. 3 in "Galaxies and the Universe" (Vol. 9 of "Stars and Stellar Systems"), Ed. A. Sandage, M. Sandage and J. Kristian, University Chicago Press, pp. 81.
Muller, C.A. and Oort, J.H. 1951, Nature 168, pp. 357.
Oort, J.H. 1927, Bull. Astron. Inst. Netherlands 3, pp. 275.
Oort, J.H. 1928, Bull. Astron. Inst. Netherlands 4, pp. 269.
Oort, J.H. 1932, Bull. Astron. Inst. Netherlands 6, pp. 249.
Oort, J.H. 1940, Astrophys. J. 91, pp. 273.
Oort, J.H. 1941, Bull. Astron. Inst. Netherlands 9, pp. 193.
Oort, J.H. 1946, Monthly Notices Roy. Astron. Soc. 106, pp. 159.
Oort, J.H. 1958, Inst. Internat. Phys. Solvay, Conseil de Phys. no. 11, pp. 163.
Oort, J.H. 1960, Bull. Astron. Inst. Netherlands 15, pp. 45.
Oort, J.H. 1962, in "Interstellar Matter in Galaxies", Ed. L. Woltjer, Benjamin, New York, pp. 234.
Oort, J.H. 1966, Bull. Astron. Inst. Netherlands 18, pp. 421.
Oort, J.H. 1967, in "Radio Astronomy and the Galactic Systems", Ed. H. van Woerden, IAU Symposium no. 31, pp. 279.
Oort, J.H. 1969, Nature 224, pp. 1158.
Oort, J.H. 1970, Astron. Astrophys. 7, pp. 381.
Oort, J.H. 1971, in "Nuclei of Galaxies", Ed. D.J.K. O'Connell, North-Holland Publ., Amsterdam, pp. 321.
Oort, J.H. 1974, in "The Formation of Galaxies", Ed. J.R. Shakeshaft, IAU Symposium no. 58, pp. 375.
Oort, J.H. 1975, in "Structure and Evolution of Galaxies", Ed. G. Setti, Reidel, Dordrecht, pp. 85.
Oort, J.H. 1976, Pub. Astron. Soc. Pacific 88, pp. 593.
Oort, J.H. 1977a, Ann. Rev. Astron. Astrophys. 15, pp. 295.
Oort, J.H. 1977b, Comments Astrophys. 7, pp. 51.
Oort, J.H., Kerr, F.J. and Westerhout, G. 1958, Monthly Notices Roy. Astron. Soc. 118, pp. 379.
Roberts, M.S. 1975, Chap. 9 in "Galaxies and the Universe" (Vol. 9 of "Stars and Stellar Systems"), Ed. A. Sandage, M. Sandage and J. Kristian, Univ. Chicago Press, pp. 309.
Rougoor, G.W. and Oort, J.H. 1960, Proc. Nat. Acad. Sci. Washington 46, pp. 1.
Rubin, V.C., Ford, W.K., Peterson, C.J. and Oort, J.H. 1977, Astrophys. J. 211, pp. 693.
Schwarzschild, M. 1954, Astron. J. 59, pp. 273.
Van Houten, C.J., Oort, J.H. and Hiltner, W.A. 1954, Astrophys. J. 120, pp. 439.
Van de Hulst, H.C., Raimond, E. and Van Woerden, H. 1957, Bull. Astron. Inst. Netherlands 14, pp. 1.
Van der Kruit, P.C. 1970, Astron. Astrophys. 4, pp. 462.
Van der Kruit, P.C. and Allen, R.J. 1978, Ann. Rev. Astron. Astrophys. 16, pp. 103.
Van der Kruit, P.C., Oort, J.H. and Mathewson, D.S. 1972, Astron. Astrophys. 21, pp. 169.
Van Woerden, H., Rougoor, G.W. and Oort, J.H. 1957, C.R. Acad. Sci. Paris 244, pp. 1691.

E. Margaret Burbidge studied at University College, London, Geoffrey R. Burbidge at the Universities of Bristol and London. The Burbidges worked at London Observatory, Yerkes Observatory, Cavendish Laboratory, Mount Wilson and Palomar Observatories, University of Chicago, and University of California at San Diego. Margaret Burbidge has been Director of the Royal Greenwich Observatory, and is now Director of the Center for Astrophysics and Space Sciences, University of California, San Diego. Geoffrey Burbidge is now Director of Kitt Peak National Observatory.

E. Margaret Burbidge

Geoffrey Burbidge

E. Margaret Burbidge, Jan H. Oort and L.V. Mirzoyan at IAU Symposium No. 29, "Non-Stable Phenomena in Galaxies", Byurakan, Armenia, U.S.S.R., May 1966. (During this Symposium Oort was taken on an excursion to see the wild tulips of Armenia, the ancestors of the Dutch tulips.)

BIRTHDAY WISHES

Sad the walk past dear Nassau roofs and Mercer elms,
As we think of Princeton's crown plucked of great jewels:
Russell of the diagram and Hermann Weyl,
And Bohr and Einstein of the quantum quest,
And many a long friend now lost in night.
But when we think of you
And your warm companion of shared hope and care,
When we think of your expeditions
To the flaming ramparts of the world,
To Crab and matter density,
To galaxy's fission and nucleus and black hole,
Our hearts leap with joy that you are with us,
Lovers of Shakespeare loved by us all.

John and Janette Wheeler visiting Princeton on your 80th birthday.

H. van Woerden, W. N. Brouw, and H. C. van de Hulst (eds.), Oort and the Universe, 151.

THE EARTH AND THE UNIVERSE

Abraham H. Oort

Down how many roads among the stars must man propel himself in search of the final secret ? The journey is difficult, immense, at times impossible, yet that will not deter some of us from attempting it. We cannot know all that has happened in the past, or the reason for all of these events, any more than we can with surety discern what lies ahead. We have joined the caravan, you might say, at a certain point; we will travel as far as we can, but we cannot in one lifetime see all that we would like to see or learn all that we hunger to know.

From the "Immense Journey" by the late anthropologist-writer Loren Eiseley.

To me, neither a colleague nor a direct friend of Jan Oort, it seemed a difficult task to write an essay in this *Liber Amicorum Astronomicorum* on the occasion of his 80th birthday. However, the unique position of being his son, and also working in a related scientific field, made me accept the invitation.

The approach I have taken is to look into my own work and life habits and, through a direct comparison with my father's habits, try to understand the style chosen by him.

One of the early impressions of my father's life was that a fuzzy boundary divided his times of "work" and "no-work". Only one door separated the office in the Sterrewacht from the warmth of our home. Transgressions occurred at times when we brought tea or coffee to his room, or when guests joined us for a meal. The boundary between work and play appeared as a thin surface, continuously changing, sometimes engulfing our living space, and then again retreating into its own sphere. It was clear that astronomy required a lot of serious, hard work and sacrifice. Perhaps no wonder that I was hesitant to enter the field of science. Nevertheless, since at the time I had no specific professional interests or outstanding abilities, pure elimination brought me to the study of science at Leiden University. Inspiration, as my father found it through

H. van Woerden, W. N. Brouw, and H. C. van de Hulst (eds.), Oort and the Universe, 153–156.

Kapteyn in Groningen, came to me years later when I entered the side field of meteorology. It took place during a two-year stay at the Massachusetts Institute of Technology (MIT) in Cambridge, Massachusetts. My advisor at MIT, Victor P. Starr, one of the greatest geophysicists of that time, finally awakened in me the potential to do original research, and gave me the necessary confidence to embark on my own. Since that time in a slowly, but steadily rising line I finally reached the frontiers of science, experiencing now the exhilaration my father must have felt all along in his career.

As far as working habits are concerned, the similarities between my father and myself are many. I may mention the need for peace and tranquillity to work in, the hate for intruding man-made noises from radio or television, the slow and sometimes painful transition either to immerse into the world of science or to come back from it into the outer, actual world. My father and I share a similar love of nature as a place for contemplation, and both feel a need for physical exercise, be it hiking or skating. Writing, the backbone of science, seems to come equally hard to both of us. Struggling with each word, writing and rewriting, crossing-out and later restoring, is a typical pattern before we are finally satisfied with what appears on paper.

Beyond this first, outer shell, the correspondence between us is less. Beside his being more talented in mathematics and theoretical comprehension, the perhaps most important difference lies in the extraordinary intensity of my father's drive to probe nature. Proceeding still deeper to the inner core, our greatest asset, intuition, may again be similar. Both of us seem to thrive on intuition and the ability to recognize, in the continuous flow of often dull and routine research, when something really important appears, both in our own work and in that of others. Once a hint of something special is found, we have the same doggedness to pursue the scent, and follow the trail to a new property of nature. On a more technical level, my strength lies in the interpretation of direct observational material, often in statistical terms, and in finding the implications thereof.

The closest parallel to my objectives in meteorology and oceanography may be found in the work of the 16th-century Danish astronomer, Tycho Brahe. He provided the observational basis for the planetary dynamics developed later by Johannes Kepler. My aim is to work toward a clear and firm observational foundation of the global atmospheric and oceanic circulations. Thus, when in the near future a new Kepler will appear, a more satisfactory and complete understanding of the operation of the earth's climate may be obtained. My father is probably more ambitious than I, combining some of the elements of both Tycho Brahe and Johannes Kepler in his pursuit of understanding the steady and nonsteady phenomena on the scale of galaxies and the universe as a whole.

In our own time, I am tempted to draw a comparison of our relationship with that between another well-known scientist, Freeman Dyson, and his son, George. Dyson, a theoretical physicist-astronomer-space engineer-biologist, and author of a great autobiography, "Disturbing the Universe", is connected as Professor of Natural Sciences with the Institute for Advanced Study in Princeton, New Jersey. Despite obvious differences, he reminds me of my father. Both tend to choose the unorthodox road, and

Photograph by courtesy of the University of Dundee

"SPIRAL GALAXY ON EARTH"

The old hurricane *Flossie* over the North Sea, 16 September 1978, from "Weather", Vol. 34, No. 8, August 1979.

to pursue it further than most other people would dare to do. Both have probably an equally strong interest in finding out what lies beyond signs of "Verboden Toegang" (No Entrance). George Dyson, in a way as unconventional as his father, chose a quite different style of life, retreating into the wilderness, living and working in nature off the deserted and wild coast of Western Canada and Alaska. George's originality was expressed in the design and construction of a big, sea-worthy canoe to explore the Pacific coast. The visit of Freeman Dyson to his son, after many years of separation, is beautifully described by Kenneth Brouwer in his book "The Starship and the Canoe".

To some extent, a similar separation has occurred in our case. I chose to go into the earth sciences rather than into the more etheric astronomy. Like George Dyson, I moved westward, emigrating with my wife and children, to a new country to live our own lives, away from family and tradition. The separation has lessened in recent years, both because of my scientific growth and through the frequent visits of my parents to Princeton. This trend, I hope, will continue and lead to a higher level of mutual respect, understanding and love in the years to come.

Abraham Oort, Jan Oort's son, studied physics at Leiden, and is now with the Environmental Research Laboratories at Princeton. He got his doctor's degree at Utrecht.

Abraham H. Oort

THE CHALLENGE OF JAN OORT

J.H. Bannier

Sitting down at my desk to start writing my contribution to the book a number of his friends are going to dedicate to Jan Oort on the occasion of his 80th birthday, I suppose that I do so almost exactly a year after he himself started to do likewise for my own 70th birthday, which fell in the middle of last year. Is this purely accidental, or is it, as I rather think, typical for the way in which our collaboration, ever since the end of World War II, has been functioning? For this has always been a matter of Jan having new ideas, taking a new initiative and putting this in front of me in such a challenging way that I was trapped, even before I knew it, into helping him. Not that I minded; for I'm sure that my own career as a Science Administrator has been thriving on this, and similar, challenges.

Even in the article he wrote for me last year, after giving a masterly summary of the developments in astronomy in the last thirty years, he could - as he wrote - not resist the temptation to put a direct question to me. He explained how a request by the Editor of the ESO Messenger to describe what he would do if ten observing nights would be allocated to him with the Very Large Telescope - so far only tentatively and loosely planned -, had inspired him to a voyage of discovery to the Middle Ages of the Universe, a time of pomp and splendour just as stirring as our Middle Ages: the time of the birth of great clusters of nebulae, a gigantic fireworks of young star systems with newly born quasars in their centre. After this introduction he asked me what we would do if we were given the chance to start three new projects, either in a national, i.e. Dutch, a European or perhaps even worldwide context: (1) a new Westerbork type radiotelescope in the Southern hemisphere, or (2) a Very Large Telescope (VLT) as he had envisaged for his imaginary voyage for the "Messenger", or (3) a project to study, with the help of artificial satellites, the nuclei of quasars and star systems with a resolution ten times better then the 0".0001 reached by the existing Very Large Baseline Interferometers.

This question so much intrigued me that I just had to answer it, which in its turn started me to answer also the other authors of that treasured booklet. As always, Jan, with only a few words, had provided me with a task that took me a considerable time to perform.

H. van Woerden, W. N. Brouw, and H. C. van de Hulst (eds.), Oort and the Universe, 157–160.

But having quoted the question I am bound also to mention what I answered. Translated, my argument ran more or less as follows: "My immediate answer: If we could do it together, I would start with the greatest possible enthusiasm on whichever of the three projects would be chosen. But I suppose you would like to know which of the three I would select. This is not a quite fair question as I - even after reading your article-just lack the scientific insight necessary for such a choice. But I would start by trying to let the scientific promise of each of the projects be compared with what appears to be financially and organisationally feasible. It is clear that a second WSRT in the Southern Hemisphere is the easiest and the cheapest, as it would not require a new design, while (if Chile would be politically acceptable) sufficient flat ground is available between La Silla and the Interamericano. Could this project be carried out in the framework of ESO, I see no insurmountable financial difficulties. To be paid for by the Netherlands alone would be another matter, and, anyhow, the project would have to be weighed against not only our expenditure on Astronomy as a whole, but also against the consequences for other fields of research. This comparison would even be more necessary in the case of the two other projects, but there also the organisational problems come in. Would a VLT be possible in the framework of ESO, then these problems would not be too large, but for your Satellite Project they would be enormous. Here the challenge would be the greatest and that, of course, is the reason why we - were we only a few years younger - should select this one! Let's hope our successors can see this in the same light."

What I have written above may have made clear that in my relations with Jan Oort "challenge" has quite often been the key word. Hence its appearance in the title of this Article, although the Editors proposed to me another title: "Motor and promotor: Oort as research initiator...". But is there such a thing as a silent motor? And the quieter Oort is, the greater is the impression he makes. A late President of Leiden University, himself an Astronomer turned Administrator, is reputed to have said: He needs only to whisper and he gets what he wants. This is indeed the way in which he set forward to get Government support for the Dwingeloo (and later the Westerbork) radiotelescope. A friendly word here and a small prod there, together with a superb feeling for finding the right people, able, willing and persevering enough to do the spade work, have enabled him not only to reach these goals but also to make lasting friendships in the process.

In 1953, Oort asked the then most prominent astronomers of Western Europe to come together in Leiden to consider with him the possibility of creating a European Observatory in the Southern Hemisphere. Every reader of these lines will know that this meeting has resulted in the European Southern Observatory (ESO) which is now established in Chile, but few will realize how difficult it was to reach that goal. One country, although originally willing to participate, withdrew in favour of a plan to build a national telescope under the Southern sky. In others it was feared that the large amounts of money needed for this international venture would drain the funds necessary for the smaller observatories of the universities and thereby endanger the creation of a new generation of astronomers. It was therefore not uncommon that heated discussions occurred in the many meetings, held all over Europe in the years between 1953 and the signing of the ESO Convention in 1962. I remember several occasions, particularly one in that rather sombre, high-vaulted meeting room over the front entrance of the Observatoire de Paris, on which Jan Oort, as always Chairman, in order to avoid a direct confrontation with the Committee member who had just made a critical remark, sat silent for several minutes, so embarrassingly long that one of the others present had to intervene and lift the spell under which we all had fallen. I have never dared to ask whether at such occasions Jan really did not know how to continue or if he did it on purpose. Certain is that his silences could be very effective.

Now it cannot be denied that our host in Paris, Professor André Danjon, could be tantalizing, like on the occasion when we were drafting the ESO Convention on the model of the CERN Convention. Fortunately the Convention ruling the European Organization for Nuclear Research (CERN) was written in French as well as English, which meant that we started with an officially approved French text, whereas the ESO Convention would be in French only. When we came to the Article regulating the financial contributions of the member states, we found that for CERN these should be proportional to the respective Gross National Incomes "at factor cost", or in French "au coût des facteurs". I'm quite sure none of the physicists who had helped formulating the CERN Convention, let alone any astronomer on the ESO Committee, understood what the economists mean when they use this expression, but no one objected to what is evidently a standard technical expression, except Professor Danjon, who was adamant that 'au coût des facteurs" is <u>not</u> French and should <u>not</u> be allowed in an official French text. (Incidentally: at this stage, the German representative asked: "Wie viel kosten hier die Briefträger?") But Danjon had his way. By referring in the ESO Convention simply to the relevant Article of the CERN Convention inclusion of the objectionable words was circumvented. And, of course, Jan Oort never showed any amazement or lack of understanding, a quality for which I have often envied and admired him.

It is therefore with considerable embarrassment that I now have to make a confession. A few years ago Dutch Television, under the title "Markant", produced a series of Portraits of well-known Netherlanders, one of them, of course, being Jan Oort. The reporter making this programme could ask rather personal questions, and while interviewing me about Jan he asked me point blank: "Is he indeed as modest as he seems to be?", to which I immediately answered: "No, he isn't".

For weeks and months I have been fearing what Jan's reaction would be when suddenly and publicly confronted with these words of mine. Fortunately it never happened. Even television producers sometimes bite more than they can swallow, and have to cut in their "takes". Although I figured in the program, this exchange of words did not appear, but nevertheless I would now like to explain to Jan why I gave this answer. I simply do not hold with modesty. If a man (or a woman, for that matter) knows that he has a sound idea, that he has something to offer, he (or she) should say so openly and without restraint. If Jan Oort had been modest, he would never have realized how much he could contribute to

Astronomy and to Science in general and he would not have known that he had really sound ideas, i.e. if he would not really have known what he wanted, then he would have kept his mouth shut. And Astronomy as well as the Netherlands would have been the poorer for that.

J.H. Bannier has been director of the Netherlands Organization for the Advancement of Pure Research (Z.W.O.) from 1949 until his retirement in 1970.

JAN OORT AT THE TELESCOPE

Fjeda Walraven

More than 30 years have passed since I first arrived in Leiden in 1946. There are many things that came to mind on this occasion of your eightieth birthday, Jan, and they might well fill a book of their own.

Where shall I start? With your support and counsel, ready whenever I needed it? With your intense interest in everything I undertook? Your help in financing and realizing projects I proposed? Or your deep interest in practical astronomy?

As an observational astronomer, I find the latter most appropriate. Often, when my wife and I were observing RR Lyrae with our first photoelectric photometer at the Zunderman reflector, your head would appear through the floor of our observing room, and you would come to show your interest in our results.

Our work on RR Lyrae occurred in that warm summer of 1947. However, your interest defied wind and cold, as became obvious when we were observing the Crab Nebula at the photographic refractor, which we had equipped with a photo-electric polarimeter. Busy though you were in the daytime, you often helped us observing at night. One of these nights it was cold enough to make the ink freeze in our Honeywell-Brown pen recorder. But, as if by magic, in a few moments you produced a hair-dryer, which served to melt the ink and help us carry on. However, on turning back to the telescope, I could not find the Crab in the eyepiece any longer. It took a while before you discovered that the lenses of the telescope had been covered with snow!

On one of those occasions, you said with a sigh that you would love to do more practical astronomy, if only the Observatory's daily affairs would allow you more time.

However, I wish to state here that your share in practical astronomy, albeit indirect, was and is of very great importance. Your deep interest in observational astronomy and counsel to practical astronomers made Leiden Observatory into what it is today.

Fjeda Walraven is associate professor at Leiden Observatory. He has been, most of his time, at the Leiden Southern Station at Hartebeespoortdam.

H. van Woerden, W. N. Brouw, and H. C. van de Hulst (eds.), Oort and the Universe, 161.

PERSONAL RECOLLECTIONS

Gart Westerhout

To Jan Oort, on his eightieth birthday.

Dear Jan,

To my great regret, I cannot attend the celebration of your eightieth birthday to convey my wishes in person. All the same, this does not make these wishes less heartfelt. I will restrict myself here to wishing you continued good health; and I will recall some memories of the past. As I have told you on another occasion, my position as Scientific Director of the U.S. Naval Observatory was very much due to your enormous influence on my career, dating back to the time that I was a young student. Your restrained enthousiasm, your methodical approach to the most complex subjects, the emphasis you always put on observations and the need for their high quality, and finally, your deep insight into the foundations of astronomy have been a tremendous stimulus and guideline for me.

Since my years as student in Leiden, especially the years as a doctoral student, I have looked back to the Oort lectures with great pleasure. During those formative years I was very impressed by the systematic manner in which you approached a problem, and emphasized the interconnection of the many facets of astronomy. In your lectures on Dynamics, Stellar Systems, Statistical Astronomy, etc., it became repeatedly obvious how the fundamental measurements influence more or less the entire field we call astronomy, and even how most of modern astronomy and astrophysics would simply not have been possible without the continued efforts in fundamental astrometry. I remember well your tremendous interest in the N30 and the later FK2, 3 and 4 fundamental catalogues, and your interest in and the continued emphasis you placed on the importance of Transit Circle observations. Was it not you who was the driving force behind the meeting in Cincinnati, some time in the fifties, which was called to discuss fundamental measurements in the Southern hemisphere? The emphasis you always put on the importance of proper motions in the study of the Galactic System greatly impressed me. As a result of this influence you have had on my astronomical interests,

H. van Woerden, W. N. Brouw, and H. C. van de Hulst (eds.), Oort and the Universe, 163–164.

I have followed closely that part of astronomy we call "fundamental", and have attempted - admittedly with very little success - to share some of this interest with my students. Since 1962, while at the University of Maryland, the U.S. Naval Observatory was a next-door neighbour, giving me the opportunity to continue discussing some of these problems and to keep more or less informed. I feel very fortunate indeed to now be so closely associated, thanks to your influence, with this most fundamental area of astronomy.

Let me now recall two anecdotes dating back 25 - 30 years. The first one was the most impressive experience I have ever had; the second is an example of what your poor students had to put up with!

In 1952, when Walraven and I, using a small portable telescope, were investigating Hartebeestpoortdam as a possible observing site - should the Leiden Southern Station be moved there or not? - you visited the Southern hemisphere for the first time. We were making measurements in the pitch dark, always somewhat concerned about the baboons wandering around the camp. At a certain moment, Jan Oort had disappeared. That is a natural event, but after an absence of 15 minutes we became concerned. We started searching, fearing you were half eaten by a baboon, lion or similar wild animal. After some considerable time, we found you on the other side of a small hill ("your flashlights are too bright"), flat on your back in the wet grass, risking pneumonia, with the Centre of the Milky Way in the zenith. You could not be convinced to get up, and you shooed us off! I have never forgotten the impression this event made on me. Here was the man who was the first to unravel the structure of the Galactic System, twenty five years earlier, and who now saw it for the first time, as a natural phenomenon, of which man is a part. Your fascination, and the theories that must have formed in your mind at that time, almost physically radiated from you.

The second anecdote is more down-to-earth: the Observatory skating tours. There one was, as a student, trying to keep up with the easy skating style of Jan Oort while at the same time paying attention to his attractive daughter. And then he would come and skate beside you, starting an intensive discussion on an astronomical subject. Just imagine the challenge: avoiding getting stuck in the rifts in the ice and in the pitfalls of the scientific problem, while at the same time trying to keep up with the speed of skates and thoughts. And if all of that worked out, the tour had been a success and one was dead tired. Not Jan Oort, however, who would then like to make a "little" detour to end up closer to home, often over the thinnest ice on earth.

Jan and Mieke, my best wishes on this day for a happy and fruitful ninth decade. Judith joins me in these wishes.

Gart Westerhout

Gart Westerhout studied at Leiden, 1945-1954, and obtained his doctorate there in 1958. In 1962 he became Professor and Director of Astronomy at the University of Maryland, where he stayed until 1977. Since 1977 he is the Scientific Director of the United States Naval Observatory in Washington, D.C., U.S.A.

STYLE OF RESEARCH

Henk van de Hulst

We all know about the existence of great scientists and we have also learned to accept the fact that some are greater than others. But why is this so ? Did they have the luck to strike gold ? Were they more alert to seize new opportunities ? Did they have a deeper insight and intelligence ? Or should we rather seek the extra factor in character and attitude ?

All these questions spring to my mind, now that I have undertaken to write a little essay on "Style of Research" at the occasion of Jan Oort's eightieth birthday.

Evidently, I shall not be able to answer these questions. All elements are present, and it could well be imagined that a set of questions of this type could be taken as a matrix in an organized attempt to characterize the style of research of a number of great scientists. At the same time such an attempt would be next to impossible, for where are the measurable quantities ? The printed word remains but forms only a condensate of what actually happened. Quantities that are derived from that condensate, like citation indexes, necessarily have the same very superficial character. An index measuring elation and frustration during the research effort would be highly relevant but is not available. Even data that can be quantized in principle, like number of letters written, number of meetings attended, number of times late for meals, or number of wastepaper baskets filled in a year, are irretrievable.

The alternate road to discuss the subject is by personal observation and comment. This road I shall try to follow, even though it is based on the hazardous material of incomplete memories.

I do not know when I met Oort first. It may have been in the early forties at one of the meetings of the Nederlandse Astronomen Club, a kind of interuniversity colloquium. But the person who comes forward if I think of those meetings is Pannekoek, who was impressive by his personality, his persuasion and his indefatigable attention to equation after equation, a person easy to identify with for a theoretical student. It may also have been during the first interuniversity summer conference of astronomy students and staff held in 1941.

The more vivid memories are of two occasions. The order does not really matter, but the first must have taken place between my participa-

H. van Woerden, W. N. Brouw, and H. C. van de Hulst (eds.), Oort and the Universe, 165–172.

tion in a prize essay of Leiden University (deadline May 1942) and what we call the "hongerwinter" (the last 6 months of war, 1944-45).

Oort visited me one day at Utrecht, en route from the place where he was staying as a precaution, to Leiden, where he visited from time to time in secrecy. The necessity to hide arose from the fact that he had been one of the vocal group protesting the dismissal of Jewish professors and forcing the closure of the University. How he managed to have a flat tire on his bicycle at such a convenient place, half-way the total 120-km route, I do not know. We sat in my father's study discussing ways in which the interstellar particles could be kept from growing too big. By that time we were convinced that the particles were too warm (10 - 20 K) to retain the hydrogen, but it seemed also unavoidable that all of the heavier atoms hitting a particle must freeze down with as much hydrogen as they could chemically bind. With the knowledge of interstellar densities and velocities available, this meant that any grain that had nucleated would grow to the average size in about a million years, at least a factor thousand short of the galactic time scale. Oort sat there, trying out on me the various catastrophies that might happen to a grain after this short time, and this later led to our joint paper on colliding clouds. Later we went into the garden to repair the flat tire. This, of course, was a very common job and every Dutch boy knew you must let the glue dry for ten minutes before you press the seal on the tube. Yet I had never seen anyone who actually had the patience and confidence to wait that long. Oort did, which impressed me more than his timing of interstellar collisions.

In a second view I come into Oort's office in Leiden. The first impression is that the room is almost dark, for a heavily overgrown verandah (the word porch is not lofty enough) filters the daylight. The room is high. It is part of the original building of about 1860 and at one time was part of De Sitter's residence. There are several objects of art, some flowers, a striking portrait of De Sitter, and further desks and bookcases of various sizes and styles. But the focus is a small wooden table with a lit desk lamp, in a nook away from the daylight side of the room. That is where Jan Oort is doing his current work and where he retires when other duties permit. There must have been an exchange of greetings when I come in, and a reminder of the topic we are going to discuss. Yet, strangely enough, my memory is mostly of silence. Oort takes a scientific problem like hot food: one bite at a time, lets it cool down, chews it, swallows it and digests it, and only much later tells how it tastes. In my education I have learned the rule not to interrupt other people in their speech. This is my first and vivid experience of the more advanced rule not to interrupt other persons in their train of thought.

The few lectures of Oort I attended as a student during a three month visit to Leiden in 1944, were a little disappointing. The setting was strange: half a dozen students and a solitary blackboard in the darkroom under the astrograph telescope, a place that even the director Hertzsprung was not supposed to know. Oort was well prepared and spoke softly from scribbled notes. The subject was galactic dynamics, and I do not know if circumstances ever permitted the course to be finished. What I remember was not a grand scheme, but tables with many columns of values of proper motions and equally many values of mean errors, a clear message that the careful assessment of observational data is the root of all astronomy.

* * *

While the preceding episodes contain some clues, they do not bring us much closer to an analysis of the general question of style of research. I return to this question now and shall broaden the subject by examining some differences of approach which I have observed among scientific research workers generally, and which might serve as elements in a typology of style. This discussion is necessarily brief and spotty. The very intriguing question how these distinctions correlate with each other must be left almost completely aside.

A. Talker or thinker ? The thinking type was sketched above. The talking type, who pours forth with comments, hints and suggestions, hoping that a few of them may prove helpful or correct, is so familiar that every reader can fill in an example. It is not altogether bad. Brainstorms are sometimes quite useful and cannot be conducted in silence.

B. Observer or theorist ? Many academic debates, both at students' and at professors' level, have been devoted to the distinction of theory and observation. For practical reasons the distinction may be useful, for it is impossible to do everything at the same time. Yet I consider both attitudes as limitations of the true scientific attitude. To adopt such a limitation temporarily is helpful as part of a methodical approach. In order to make truly good observations, it is useful to drop the bias of theoretical expectation. Conversely, it is often necessary to say: never mind the observations; I first wish to explore the precise consequences of a particular set of assumptions. Many scientists find the right rhythm to hop back and forth between these attitudes, and I have the feeling that this art is more common now than it used to be.

C. Generalist or specialist ? Any researcher finds sooner or later that he cannot do or follow every subject within his competence. Some choose radically a certain well-defined claim, often mark it by writing a book or two, and remain a specialist on that subject during the course of their active life. Others are content with a less complete (but not necessarily less profound) knowledge on a much broader field and remain generalists. Often they lead a somewhat nomadic existence, seeking new green pastures, or returning to old ones after many years.

D. Pioneer or consolidator ? Think of finding a road through difficult terrain. Someone has to find the first trail, which may be winding, unmarked and hazardous but reaches the goal. He is the pioneer. However, if this road is to be used for heavy traffic or transport, which is often an essential preparation even for further pioneering efforts, more is necessary. The consolidators come, cut curves, remove hazards, erect road signs, and make a bridge where there was only a ford. To translate this metaphor into science does not need further elaboration.

E. Researcher or teacher ? The choice between these alternatives may be decided by job opportunities rather than by personal inclination. But among those of us who are in the privileged position to be able to perform both functions, preferences for one or the other may be observed. The pleasure to make quite sure, by no matter what devious route, how the actual Universe is arranged, is a typical researcher's pleasure. The pleasure to explain phenomena and relations in the clearest and simplest manner is typically the teacher's. And the temptation to select the facts or tamper with them for this purpose (We are told Ptolemy did so already) is always there.

F. Position defender or problem solver ? This is a tricky distinction which has puzzled me a great deal. The simplest way to explain the distinction is in terms of the chess game. In a normal chess game each person has one side. He has to improve or defend his own position, and in the course of doing so he can gamble or bluff as he sees fit, for the duty to find the best moves of the opponent does not rest on him. The task and attitude in chess-problem solving is radically different. The sentence "White mates in three moves" has the unwritten sequel "even in the presence of the best counter moves". Consequently, the problem-solver has to examine the very best possibilities for both sides. The existence of analogous differences in the attitudes toward scientific research is striking to anyone who opens his attention to this aspect. My impression is that the attitudes are adopted on very diverse grounds. To some extent the choice may be a matter of character. Most people find it easier and more gratifying to defend "their own" point of view. Economic factors may also be involved. I fear that an undue emphasis on personal credit fosters the scoring of personal points, and thus forces young scientists to take "their own" side. This happens notably in the U.S.A., but more recently also in Europe. A third and probably deeper basis for choice is in the cultural tradition. The French have a proverb that truth emerges from the clash of opinions, and in a number of British universities the tradition to adopt and defend extreme positions is strong. That this is not an accidental feature is shown by the fact that also in legal pleas the British system permits stronger arguments starting with "I submit" than the continental system does. The tradition in which I have been brought up is that in principle there are no sides, and that the salomonic attitude of judging right without bias is always the best.

G. Individualist or team worker ? This distinction is largely one of character. Some people feel happier alone, and others while working in a team. Yet it is also connected with the technical requirements of research. The advanced systems necessary in major research efforts today require a team effort in which a research worker is one of many. He has to place confidence in the competent handling of other subsystems by someone else on the team. If he somehow cannot live with this set-up, his choise of research topics is extremely limited.

H. Perfectionist or ....? I fail to find the antonym, but this distinction is one of the most important of all, for one of the strongest temptations of a scientist is to become a perfectionist and thereby see the vitality of his research effort thwarted. I have heard many practical physicists and astronomers complain about the mechanical engineers, who never construct the simple gadget that was requested but manage to come up with a glossy instrument, thus wasting time and money. They usually forget that in their own research work they are subject to exactly the same temptation. Many a scientist has spent part of a life time perfectioning a tool, be it theoretical or practical, without checking that the customer for that tool was still anxiously waiting.

* * *

I have presented the preceding list of contrasting attitudes in a brief form. Just enough comments were added to reveal my own preferences (it would have been impossible to conceal them),but no illustrations or applications to a particular field of research or to particular persons were included in this description. We shall now return to the original questions and examine which of these attitudes are conducive to research of excellent quality. At the same time, I shall have the occasion to mention which characteristics appear most strongly in Oort's style of work,according to my limited observation.

If I skip the normal hazards of declining health, declining intelligence, or declining spirit, I see perfectionism as the most severe danger to excellent research. It starts well: many discoveries both in astronomy and in other sciences came to pass as a consequence of the stubborn effort to improve the instrumental precision. The same is true for theoretical work. Where would Einstein's test of general relativity from the precession of Mercury's perihelium have been, if not preceding generations of researchers had patiently unraveled the 130 times larger effect due to perturbations by the planets ? However, somewhere a point of diminishing returns is reached. Many people yield to the temptation to go beyond that point, for the work remains equally exciting to their own taste. Yet, how silly would it look if someone now continued to put all his effort for this purpose into the motion of Mercury, unaware of the fact that since 5 years we have a binary pulsar showing the same effect on a scale more than 300,000 times larger.

This reveals the best antidote against perfectionism: to remain open to and get inspired by an even more exciting problem. Oort manages to do so quite well (I do not have to use past tense) and this point alone, a very genuine interest in the marvellous Universe, may very well be the key to his style of research.

The problem that comes next in difficulty is to choose the proper measure of specialization. Both a too narrow specialization and a too diffuse general interest detract from optimum results. Clearly, the choice depends not only on inclination but also on personal capacity. Oort once said: "you can do only one thing well at a time". I think he follows this maxim himself. But he has managed to do many things well at different times. In this respect his style of work must be classed as somewhat nomadic, although he returns very frequently to his preferred pasture, the structure of our Galaxy and its mysterious centre.

A related facet of this attitude is that Oort never used (or needed) the traditional means to solidify a claim by writing a book. He started one but did not take the time to complete it, for the researcher always jumped ahead of the teacher. This attitude is also noticeable in the way his research papers are written. There were, especially in his earlier papers, numerous additions and footnotes, even to the proof stage. All of these were directed at a better grasp of the research topic, none at a further clarification for a casual reader. In one paper an illustration was added because I insisted, like I do now with my Ph.D. students, that an average reader must be able to understand. Generally, however, he resisted the idea of polishing the presentation for didactic reasons, and justified this by saying that the best paper was one that followed faithfully the argument as it had developed in the mind of the author.

This statement sounds rather cocksure and invites some comments about the categories of modesty and ambition. On purpose I did not include those words among the list of distinctive properties that may be used to characterize style of research. I feel that a basic self-confidence fed by some form of ambition is a necessary ingredient of any researcher, no matter what style. The manifestations may be more or less subtle, but modesty truly does not exist and, therefore, cannot be used as a distinction.

Absence of personal claimstaking is something different and may open the way to good team-work. Please forget the present big-science association of "team" with organigrams and interface descriptions. Oort performs well in such a large team, as his many practical contributions to the Westerbork observing programme show, but in the present context more relevant are the teams of two. His long association with J.M. Burgers beyond the wartime conversations about the dynamics of interstellar matter, his close work with Baade and with Spitzer, are all fine examples of work in a small, truly efficient team. The same may be said about his work with many students, although for some the unrelenting pressure of ever more facets of the problem that "it would be nice" to explore has been too strong.

* * *

Perhaps the preceding analysis has left the impression that a person must be able to select his style of research at will. This is not true, for character and training determine many of the elements, and forcing them would lead to bad results. Yet it is inspiring to observe the style in which other people work, and useful to give a student an occasional hint how he can improve his.

Perhaps, also, the emphasis on a few negative points may have left the impression that the key to excellent research is the avoidance of certain hazards, notably perfectionism and claimstaking. This impression would again be wrong. For these considerations are subordinate, certainly in Jan Oort's style of work, to the real excitement about and inspiration by the actual Universe. Basically, he remains an ardent observer. Says Jan H. Oort, anno 1980 after attending a colloquium on the binary pulsar: "It is marvellous what Nature provides for us".

Henk C. van de Hulst studied at Utrecht, 1936-1946, worked two years at Yerkes Observatory and has since 1948 been Professor of Theoretical Astronomy at Leiden.

Henk van de Hulst

Oort at the First Astronomers' Summer Conference, Doorn, 5 June 1941.
(Photo courtesy J.Houtgast)

(1)

Copies van
meisen gesand
aan O.Z. en p.2

Mission to Halley's Comet

At the meeting of experts in Leiden on May 1 I have been in particular impressed by the uniqueness of the project. I had not clearly realized before that Halley's comet will be intrinsically two orders of magnitude brighter than all other comets that can be reached by a space mission for a long time to come. A similar factor applies to the degree of activity. In this respect the proposal for a fly-by mission to Halley's comet distinguishes itself from all other space projects. The fact that it is the only chance in a "life time" lays a heavy responsibility on those who have to decide.

Another thing which I had not previously considered is that the expected optical depth of the dust close to the solid nucleus is expected to be small, so that it may be possible to observe the centres of gas and dust expulsion right down to the surface of the nucleus. Judging from the observations in 1910 (which were extensively, almost daily collected and discussed) (alinea)

The pictures that can be taken during a fly-by of comet Halley may therefore be spectacular.

by N.T. Bobrovnikoff in Publ. Lick Observatory 17, Pt I (1931),

that came forward at this meeting, and

(2)

Halley's comet showed during the three weeks it was between 0.6 and 1.0 A.U. from the Sun and between ~0.3 and 0.5 A.U. from the Earth some half dozen explosion centres each day. These produced jets, often with various condensations, moving at considerable velocities (0.1 - 1.0 km/s) relative to the comet; the condensation acted often as independent explosion centres. The accompanying Figures give illustrations of the complicated structures that sometimes showed; the overall size of each picture is about 100" (~ 100 000 km), except for Fig. 76 C which is ~ 200" long.

Probably many explosions would be observed from the spacecraft, during the week it would be within ~ 0.1 A.U. from the comet. The comet would be likely to present a spectacular appearance, and yield ample opportunities for studying in detail what happens at the explosions.

//

I have also been nearly impressed by the fact that the significance of the proposed mission extends beyond the comet itself, in as much as it may produce new insight into

the composition of the primeval "solar nebula"

the interplanetary dust particles

and possibly

the interstellar medium.

J. H. Oort
May 9, 1980

Prof. H. C. van de Hulst

9 May 1980

Dear Henk

I am enclosing some considerations which occurred to me during and after the meeting of May 1 to which Mayo had invited me.

Yours

J. H. Oort

2 illustrations uit Lick Bull. pp. 396 en 408

Comments by Jan H. Oort, 9 May 1980, on a possible space mission by the European Space Agency to Halley's comet.

BIBLIOGRAPHY OF JAN HENDRIK OORT 1922 - 1979

Some peculiarities in the motion of stars of high velocity.
B.A.N. 1, pp. 133-137, (No. 23), 1922.

The frequency of a component of the linear velocity for stars brighter than 5.8m of spectral types F, G, K and M, derived from the proper motions of Boss' Catalogue.
J. Schilt and J.H. Oort, Hoitsema Brothers, Groningen, 1923.

On the proper motions of stars of the thirteenth magnitude.
J.H. Oort and H.M. Marsh, Popular Astronomy 32, pp. 559-561, 1924a.

A comparison of the average velocity of binaries with that of single stars.
Astron. J. 35, pp. 141-144, 1924b.

Note on the difference in velocity between absolutely bright and faint stars.
Proc. Nat. Acad. Sci. Washington 10, pp. 253-256, 1924c.

On a possible relation between globular clusters and stars of high velocity.
Proc. Nat. Acad. Sci. Washington 10, pp. 256-260, 1924d.

Een discussie van declinaties uit azimuthwaarnemingen, gedaan door C. Sanders te Matuba, Portugeesch Congo.
Verslagen Kon. Ned. Akad. Wetenschappen, Afd. Natuurkunde 34, pp. 44-48, 1925a.

A discussion of the determination of declinations from azimuth measures made near the equator.
C. Sanders and J.H. Oort, B.A.N. 2, pp. 201-208 and 2, p. V, (No. 76), 1925b.

Voorlopig schema voor de bepaling van fundamenteele declinaties uit azimuthwaarnemingen.
W. de Sitter and J.H. Oort, Verslagen Kon. Ned. Akad. Wetenschappen, Afd. Natuurkunde 34, pp. 584-591, 1925c.

Provisional scheme for the determination of fundamental declinations from azimuth observations.
W. de Sitter and J.H. Oort, B.A.N. 3, pp. 1-6, (No. 81), 1925d.

De snelheid van de zon ten opzichte van de helderste sterren.
J.H. Oort and N.W. Doorn, Verslagen Kon. Ned. Akad. Wetenschappen, Afd. Natuurkunde 34, pp. 947-948, 1925e.

The solar motion from radial velocities of the absolutely brightest stars of spectra F, G, K and M.
J.H. Oort and N.W. Doorn, B.A.N. 3, pp. 71-73, (No. 89), 1925f.

Asymmetry in the distribution of stellar velocities.
The Observatory 49, pp. 302-304, 1926a.

The stars of high velocity (Doctor's thesis, Groningen University).
Publ. Kapteyn Astron. Lab. Groningen 40, pp. 1-75, 1926b.

De sterren met groote snelheid.
Hemel en Dampkring 24, pp. 338-348, 1926c.

Niet-lichtgevende materie in het sterrenstelsel (Public lecture, Leiden University, 1926).
Hemel en Dampkring 25, pp. 13-21 and pp. 60-70, 1927a.

De rotatie van het Melkwegstelsel.
Verslagen Kon. Ned. Akad. Wetenschappen, Afd. Natuurkunde 36, pp. 667-680, 1927b.

Observational evidence confirming Lindblad's hypothesis of a rotation of the Galactic System.
B.A.N. 3, pp. 275-282, (No. 120), 1927c.

Investigations concerning the rotational motion of the Galactic System, together with new determinations of secular parallaxes, precession and motion of the equinox.
B.A.N. 4, pp. 79-89 and p. 94, (No. 132), 1927d.

Additional notes concerning the rotation of the Galactic System.
B.A.N. 4, pp. 91-92, (No. 133), 1927e.

Summary of the principal radial velocity data used for the results of B.A.N. 120 and 132.
B.A.N. 4, pp. 93-94, (No. 133), 1927f.

Rotatie en bouw van het Melkwegstelsel.
De Natuur 47, pp. 238-241, 1927g.

An investigation of the constant of refraction from observations at Leiden and at the Cape.
B.A.N. 4, pp. 137-142, (No. 143), 1928a.

Dynamics of the Galactic System in the vicinity of the Sun.
B.A.N. 4, pp. 269-284, (No. 159), 1928b.

Catalogue of 460 stars for the epoch and equinox 1885.0, from meridian observations made at Leiden in the years 1880 - 1897.
Ann. Sterrewacht Leiden 13, part 4, 1928c.

Note on O. Struve's intensity estimates of the calcium K-line in early-type stars.
B.A.N. 5, pp. 105-108, (No. 177), 1929a.

Vallende sterren.
Kampvuur 3, pp. 340-342, 1929b.

The estimated number of high velocities among stars selected according to proper motion.
B.A.N. 5, pp. 189-192, (No. 189), 1930a.

The motion of the sun with respect to interstellar gas.
B.A.N. 5, pp. 192-194, (No. 189), 1930b.

Note on the velocities of extragalactic nebulae.
B.A.N. 5, pp. 239-241, (No. 196), 1930c.

De Nederlandsche Sterrekunde in de laatste halve eeuw.
Uit: De Ontwikkeling der Natuurwetenschappen in Nederland gedurende de laatste halve eeuw (samengesteld voor de wetenschappelijke afdeeling van het Ned. Paviljoen op de tentoonstelling te Luik in 1930), 1930d.

Some problems concerning the distribution of luminosities and peculiar velocities of extragalactic nebulae.
B.A.N. 6, pp. 155-160, (No. 226), 1931a.

Licht-absorptie in het Melkwegstelsel.
Hemel en Dampkring 29, pp. 41-49, 1931b.

Kan de natuurwetenschap een factor worden in het geestesleven van de moderne mensch ?
De Smidse 6, p. 1, 1931c.

Hooge School en Maatschappij.
De Smidse 6, pp. 113-118, 1931d.

The force exerted by the stellar system in the direction perpendicular to the galactic plane and some related problems.
B.A.N. 6, pp. 249-287, (No. 238), 1932a.

Note on the distribution of luminosities of K and M giants.
B.A.N. 6, pp. 289-294, (No. 239), 1932b.

Bespreking van G.P. Kuiper's dissertatie "Statistische onderzoekingen van dubbelsterren".
Hemel en Dampkring 31, pp. 307-311, 1933.

De brieven van Kepler.
De Smidse 9, p. 44, 1934a.

In memoriam Professor Dr. W. de Sitter.
Leids Universiteitsblad 4, No. 6, 1934b.

Obituary for Willem de Sitter.
The Observatory 58, pp. 22-27, 1935.

De rotatie van het Melkwegstelsel.
Natuurkundige Voordrachten Maatschappij Diligentia, Nieuwe Reeks 14 pp. 83-100, 1935/36a.

De bouw der sterrenstelsels (Inaugural lecture, Leiden University, 1935).
Hemel en Dampkring 34, pp. 1-16, 1936b.

Mean parallaxes of faint stars derived from the Radcliffe Catalogue of proper motions.
B.A.N. 8, pp. 75-104, (No. 290), 1936c.

A redetermination of the constant of precession, the motion of the equinox and the rotation of the Galaxy from faint stars observed at the McCormick Observatory.
B.A.N. 8, pp. 149-155, (No. 298), 1937.

Over de structuur van het Melkwegstelsel.
Verslagen Kon. Ned. Akad. Wetenschappen, Afd. Natuurkunde 47, pp. 39-40, 1938a.

Quelques résultats concernant la répartition de la matière interstellaire et la structure du système galactique.
Annales d'Astrophysique 1, pp. 71-96, 1938b.

Absorption and density distribution in the Galactic System.
B.A.N. 8, pp. 233-264, (No. 308), 1938c.

Evidence for spiral-like structure in our stellar system ?
Trans. IAU 6, pp. 457-458, 1938d.

Motions of RR Lyrae variables.
B.A.N. 8, pp. 337-338, (No. 318), 1939a.

De nieuwe dubbele astrograaf van de afdeeling der Leidsche Sterrewacht te Johannesburg.
Hemel en Dampkring 37, pp. 129-133, 1939b.

Review of "Stellar Dynamics" by W.M. Smart.
The Observatory 62, pp. 133-138, 1939c.

Stellar motions.
Mon. Not. Roy. Astr. Soc. 99, pp. 369-384, 1939d.

Sixth General Assembly of the International Astronomical Union, Stockholm 1938.
J.H. Oort (editor), Trans. IAU 6, pp. 1-518, 1939e.

Some problems concerning the structure and dynamics of the Galactic System and the elliptical nebulae NGC 3115 and 4494.
Astrophys. J. 91, pp. 273-306, 1940.

Intensiteitsverdeeling in extragalactische nevels.
Verslagen Kon. Ned. Akad. Wetenschappen, Afd. Natuurkunde 50, pp. 40-42, 1941a.

The attractive force of the Galactic System as determined from the distribution of RR Lyrae variables.
J.H. Oort and A.J.J. van Woerkom, B.A.N. 9, pp. 185-188, (No. 338) 1941b. (Erratum: B.A.N. 11, p. 270, 1951.)

Note on the structure of the inner parts of the Galactic System.
B.A.N. 9, pp. 193-196, (No. 338), 1941c.

Jongere Nederlandsche sterrekundigen op belangrijke posten in het buitenland.
Hemel en Dampkring 39, pp. 355-363, 1941d.

Moderne opvattingen over de structuur van het heelal.
In: Antieke en Moderne Kosmologie, Arnhem, pp. 167-181, 1941e.

Further data bearing on the identification of the Crab Nebula with the supernova of 1054 A.D.
N.U. Mayall and J.H. Oort, Publ. Astron. Soc. Pacific 54, pp. 95-104, 1942a.

A determination of the galactic pole from stars at large distances from the galactic plane.
B.A.N. 9, pp. 324-325, (No. 353), 1942b.

Note on the distances and motions of some extremely remote Cepheids in Cygnus.
J.H. Oort and P.Th. Oosterhoff, B.A.N. 9, pp. 325-327, (No. 353), 1942c.

On the relation between velocity- and density-distribution of long-period variables.
J.H. Oort and J.J.M. van Tulder, B.A.N. 9, pp. 327-331, (No. 353), 1942d.

Remark on the distances of long-period variables.
J.H. Oort and J.J.M. van Tulder, B.A.N. 9, pp. 332-334, (No. 353), 1942e.

The constants of differential rotation and the ratio of the two galactic axes of the velocity ellipsoid in the case when peculiar motions are not negligible.
B.A.N. 9, pp. 334-336, (No. 353), 1942f.

De twee sterstroomen.
Hemel en Dampkring 41, pp. 40-46, 1943a.

Some remarks on the fundamental systems of the General Catalogue and the Dritter Fundamentalkatalog.
B.A.N. 9, pp. 417-422, (No. 357), 1943b.

Tentative corrections to the FK3 and GC systems of declinations.
B.A.N. 9, pp. 423-424, (No. 357), 1943c.

The constants of precession and of galactic rotation.
B.A.N. 9, pp. 424-427, (No. 357), 1943d.

De vorming van vaste deeltjes in het interstellaire gas.
D. ter Haar, H.C. van de Hulst, J.H. Oort and A.J.J. van Woerkom, Nederl. Tijdschr. Natuurkunde 10, pp. 238-257, 1943e.

In memoriam Professor Frank Schlesinger.
Hemel en Dampkring 43, pp. 27-30, 1945.

Gas en vaste stof in de interstellaire ruimte.
J.H. Oort and H.C. van de Hulst, Verslagen Kon. Ned. Akad. Wetenschappen, Afd. Natuurkunde 55, pp. 18-19, 1946a.

Gas and smoke in interstellar space.
J.H. Oort and H.C. van de Hulst, B.A.N. 10, pp. 187-204, (No. 376), 1946b.

In memoriam Dr. A. de Sitter.
Hemel en Dampkring 44, p. 33, 1946c.

Bij het afscheid van Professor Ejnar Hertzsprung.
Hemel en Dampkring 44, pp. 166-168, 1946d.

Some phenomena connected with interstellar matter (George Darwin Lecture, 1946).
Mon. Not. Roy. Astr. Soc. 106, pp. 159-179, 1946e.

Lichtgevende gasnevels in het sterrenstelsel.
Ned. Tijdschr. Natuurkunde 12, pp. 153-155, 1946f.

Vormen en interne bewegingen van planetaire nevels.
Ned. Tijdschr. Natuurkunde 12, pp. 195-205, 1946g.

Obituary for W.Chr. Martin, A. de Sitter and J. Uitterdijk.
The Observatory 66, pp. 265-266, 1946h.

Invloed van het interstellaire medium op door sterren uitgestooten gasschillen.
Verslagen Kon. Ned. Akad. Wetenschappen, Afd. Natuurkunde 56, pp. 4-5, 1947a.

Report of the Committee of the American Astronomical Society on the merits of the proposed international project of determining accurate proper motions with the Photographic Zenith Tube.
Astron. J. 52, pp. 133-135, 1947b.

In memoriam Dr. H. van Gent.
Hemel en Dampkring 45, pp. 159-160, 1947c.

Levensbericht van A. van Maanen.
Jaarboek Kon. Ned. Akad. Wetenschappen, 1947-1948, pp. 163-167, 1948a.

Quelques remarques sur la nébuleuse autour de la Nova Persei 1901.
Comm. et Mém. des Réunions Scientifiques pour le Centenaire de la découverte de Neptune, Comité National Français d'Astronomie, pp. 11-14, 1948b.

L'orientation de l'ellipsoide de vitesses pour les étoiles faibles.
Ibidem, pp. 119-120, 1948c.

Boekbespreking van "Een Menschenleven" door Mevr. de Sitter.
Nieuwe Rotterdamse Courant, 12 febr. 1949a.

De herkomst der lang-periodieke kometen.
Verslagen Kon. Ned. Akad. Wetenschappen, Afd. Natuurkunde 58, pp. 43-45, 1949b.

Seventh General Assembly of the International Astronomical Union, Zürich 1948.
J.H. Oort (editor), Trans. IAU 7, pp. 1-552, 1950a.

The structure of the cloud of comets surrounding the solar system, and a hypothesis concerning its origin.
B.A.N. 11, pp. 91-100, and 11, p. VIII, (No. 408), 1950b.

Some remarks concerning the determination of the constant of precession.
Bull. Astronomique 15, pp. 217-228, 1950c.

Différences entre les nouvelles et les anciennes comètes.
Réunion des Astronomes des Pays Voisins, Membres de l'Union Astronomique Internationale, 1950d.

Jacobus Cornelius Kapteyn.
Hemel en Dampkring 49, pp. 1-4, 1951a.

Symposium over de werkzaamheid der Stichting Radiostraling van Zon en Melkweg en over de ontdekking van een emissie-lijn van interstellaire waterstof.
J.H. Oort, H. Rinia and M.G.J. Minnaert, Verslagen Kon. Ned. Akad. Wetenschappen, Afd. Natuurkunde 60, pp. 53-62, 1951b.

Differences between new and old comets.
J.H. Oort and M. Schmidt, B.A.N. 11, pp. 259-269, (No. 419), 1951c.

A comparison of the intensity distribution of radiofrequency radiation with a model of the Galactic System.
G. Westerhout and J.H. Oort, B.A.N. 11, pp. 323-333, (No. 426), 1951d.

A new determination of the precession and the constants of galactic rotation.
H.R. Morgan and J.H. Oort, B.A.N. 11, pp. 379-384, (No. 431), 1951e.

Introduction to IUTAM-IAU Symposium, Paris 1949.
In: "Problems of Cosmical Aerodynamics", Dayton: Central Air Documents Office, pp. 1-6, 1951f.

Interaction of nova and supernova shells with the interstellar medium.
Ibidem, pp. 118-124, 1951g.

The interstellar hydrogen line at 1.420 Mc/sec, and an estimate of galactic rotation.
C.A. Muller and J.H. Oort, Nature 168, pp. 357-358, 1951h.

Oorsprong en ontwikkeling van kometen.
Natuurkundige Voordrachten Maatschappij Diligentia, Nieuwe Reeks 30, pp. 37-48, 1951i.

Origin and development of comets (Halley Lecture, 1951).
The Observatory 71, pp. 129-144, 1951j.

The origin and nature of comets.
Proc. Math. Phys. Soc. Egypt 4, pp. 79-96, 1951k.

Spiraalstructuur van het Melkwegstelsel.
J.H. Oort, H.C. van de Hulst and C.A. Muller, Verslagen Kon. Ned. Akad. Wetenschappen, Afd. Natuurkunde 61, pp. 140-143, 1952a.

Problems of Galactic Structure (Henry Norris Russell Lecture, 1951).
Astrophys. J. 116, pp. 233-250, 1952b.

Radiostraling uit de wereldruimte.
Nederl. Tijdschr. Natuurkunde 18, pp. 116-118, 1952c.

De monochromatische emissie van de interstellaire waterstof.
Nederl. Tijdschr. Natuurkunde 18, pp. 151-154, 1952d.

Spiral structure and interstellar radio emission.
J.H. Oort and C.A. Muller, South-African Journal of Science 49, pp. 87-92, 1952e.
Monthly Notes Astr. Soc. South Africa 11, pp. 65-70, 1952f.

Expanding groups of B-stars.
Monthly Notes Astr. Soc. South Africa 11, pp. 91-92, 1952g.

L'Hydrogène interstellaire.
Ciel et Terre 59, pp. 117-133, 1953a.

De familie van het firmament; nieuwe onderzoekingen van het Melkwegstelsel.
Elseviers Weekblad, 15 Augustus, 1953b.

Spiraalvormige armen ontdekt in het Melkwegstelsel.
Hemel en Dampkring 51, pp. 153-168, 1953c.

Inferences on the origin of comets derivable from the distribution of the reciprocal major axes.
In: "La physique des comètes", Mém. Soc. Roy. Sci. Liège (4) 13, pp. 364-371, 1953d.

Oorsprong en ontwikkeling der kometen.
Tydskrif vir Wetenskap en Kuns (South Africa), pp. 142-156, 1953e.

De rotatie van het Melkwegstelsel.
J.H. Oort, K.K. Kwee, C.A. Muller and G. Westerhout, Verslagen Kon. Ned. Akad. Wetenschappen, Afd. Natuurkunde 63, pp. 94-98, 1954a.

Photoelectric measurements of extragalactic nebulae.
C.J. van Houten, J.H. Oort and W.A. Hiltner, Astrophys. J. 120, pp. 439-453, 1954b.

The spiral structure of the outer part of the Galactic System derived from the hydrogen emission at 21-cm wavelength.
H.C. van de Hulst, C.A. Muller and J.H. Oort, B.A.N. 12, pp. 117-149, (No. 452), 1954c.

Outline of a theory on the origin and acceleration of interstellar clouds and O-associations.
B.A.N. 12, pp. 177-186, (No. 455), 1954d.

Report of Commission 33 (Stellar Statistics).
Trans. IAU 8, pp. 500-511, 1954e.

Purpose and requirements for proper motions of faint stars.
Trans. IAU 8, pp. 783-784, 1954f.

Moderne Untersuchungen über die Struktur der Milchstrasse.
Die Naturwissenschaften 41, pp. 73-80, 1954g.

Special report on interstellar hydrogen.
J.H. Oort, J.L. Pawsey and E.M. Purcell, Special Report 5, U.R.S.I. pp. 47-72, 1954h.

Polarisatie van het licht van de Krabnevel.
J.H. Oort and Th. Walraven, Verslagen Kon. Ned. Akad. Wetenschappen, Afd. Natuurkunde 64, pp. 64-66, 1955a.

Acceleration of interstellar clouds by O-type stars.
J.H. Oort and L. Spitzer, Astrophys. J. 121, pp. 6-23, 1955b.

In het werk van de Sterrewacht kwam grote verandering, door invoering van electronische methoden.
Algemeen Handelsblad, 3 juni 1955c.

Information on velocity and density distribution in the interstellar gas, derived from absorption lines and 21-cm radiation.
In: "Gas Dynamics of Cosmic Clouds", Editors: J.M. Burgers and H.C. van de Hulst, IAU Symp. 2, pp. 20-26, 1955d.

Outline of a theory on the origin and acceleration of interstellar clouds and O-associations.
IAU Symp. 2, pp. 147-158, 1955e.

Solid particles in extra-galactic nebulae.
In: "Les particules solides dans les astres", Mém. Soc. Roy. Sci. Liège (4) 15, pp. 407-410, 1955f.

Measures of the 21-cm line emitted by interstellar hydrogen.
Vistas in Astronomy 1, pp. 607-616, 1955g.

Problemen betreffende de spiraalstructuur der sterrentelsels.
In: "In het Voetspoor van Kapteyn", Essays written for Van Rhijn's 70th birthday, Nederl. Astronomen Club, pp. 40-51, 1956a.

Polarization and composition of the Crab Nebula.
J.H. Oort and Th. Walraven, B.A.N. 12, pp. 285-308 (No. 462), 1956b.

Nederlands venster op het Heelal.
Elseviers Weekblad, 14 april 1956c.

Die Spiralstruktur des Milchstrassensystems.
Mitt. Astron. Gesellschaft (1955), pp. 83-87, 1956d.

The evolution of galaxies.
Scientific American 195, pp. 101-108, 1956e.

Hochfrequenzstrahlungen aus dem Weltraum.
NTZ 1, pp. 39-41, 1956/57a.

Perspectieven der Radio-Astronomie.
Nieuwe Verhandelingen Bataafsch Genootschap der Proefondervindelijke Wijsbegeerte, 3e Reeks, le deel, 4e stuk, 1957b.

Expansion d'une structure spirale dans le noyau du Système Galactique, et position de la radio-source Sagittarius A.
H. van Woerden, G.W. Rougoor and J.H. Oort, C.R. Acad. Sci. Paris 244, pp. 1691-1695, 1957c.

Polarization and the radiating mechanism of the Crab Nebula.
J.H. Oort and Th. Walraven, in: "Radio Astronomy", Ed. H.C. van de Hulst, IAU Symp. 4, pp. 197-200, 1957d.

Report of Commission 33 (Stellar Statistics).
Trans. IAU 9, pp. 476-486, 1957e.

The Crab Nebula.
Scientific American 196, pp. 52-60, 1957f.

Die neue Erschliessung des Weltalls durch die Radioastronomie.
Universitas 12, pp. 379-387, 1957g.

Het interstellaire gas in het centrale deel van het Melkwegstelsel.
J.H. Oort and G.W. Rougoor, Verslagen Kon. Ned. Akad. Wetenschappen, Afd. Natuurkunde 67, pp. 139-143, 1958a.

Bij het Internationale Congres voor Ruimtevaart te Amsterdam 1958.
Hemel en Dampkring 56, pp. 170-171, 1958b.

Comparison of the Galactic System with other stellar systems.
IAU Symp. 5, pp. 69-72, 1958c.

The Galactic System as a spiral nebula.
J.H. Oort, F.J. Kerr and G. Westerhout, Mon. Not. Roy. Astr. Soc. 118, pp. 379-389, 1958d.

Neutral hydrogen in galaxies.
In: "Stellar Populations", Ed. D.J.K. O'Connell, North Holland Publ., Amsterdam, pp. 25-30, 1958e.

Dynamics and evolution of the Galaxy, in so far as relevant to the problem of the populations.
Ibidem, pp. 415-425, 1958f.

Summary - from the astronomical point of view.
Ibidem, pp. 507-516, 1958g.

Distribution of galaxies and density in the Universe.
In: "La structure et l'évolution de l'Univers", Inst. Intern. de Physique Solvay, Onzième Conseil de Physique, pp. 163-181, 1958h.

Ruimtevaart en haar betekenis voor de natuurwetenschap.
Jaarboek Kon. Ned. Akad. Wetenschappen, 1958/1959, pp. 147-171, 1959a.

The interstellar gas in the central part of the Galaxy.
J.H. Oort and G.W. Rougoor, Astron. J. 64, pp. 130-131, 1959b.

Structure and dynamics of Messier 3.
J.H. Oort and G. van Herk, B.A.N. 14, pp. 299-321, (No.491), 1959c.

Radio-frequency studies of galactic structure.
Handbuch der Physik, Ed. S. Flügge, Springer Verlag, Berlin, Vol. 53, pp. 100-128, 1959d.

Toespraak bij 25-jarig jubileum van het Zeiss Planetarium.
Hemel en Dampkring 57, pp. 79-80, 1959e.

A summary and assessment of current 21-cm results concerning spiral and disk structures in our Galaxy.
In: "Paris Symposium on Radio Astronomy", Ed. R.N. Bracewell, IAU Symp. 9, pp. 409-415, 1959f.

Neutral hydrogen in the central part of the Galactic System.
G.W. Rougoor and J.H. Oort, IAU Symp. 9, pp. 416-422, 1959g.

La radioastronomie: Fenêtre ouverte sur l'Univers.
Journal Internat. Telecomm. Union 10, pp. 226-230, 1959h.

Internal motion and density distribution in a globular cluster.
J.H. Oort and G. van Herk, Annales d'Astrophysique 23, pp. 375-378, 1960a.

Very accurate positions of selected stars.
Astron. J. 65, pp. 229-231, 1960b.

Note on the determination of K_z and on the mass density near the Sun.
B.A.N. 15, pp. 45-53, (No. 494), 1960c.

Maatschappij en Zuivere Wetenschap.
Hemel en Dampkring 58, pp. 254-255, 1960d.

Prof. F.J.M. Stratton.
News Digest of the I.A.R.F. 44, p. 24, 1960e.

The position of the Galactic Centre.
J.H. Oort and G.W. Rougoor, Mon. Not. Roy. Astr. Soc. 121, pp. 171-173, 1960f.

Radio Astronomy - a window on the Universe.
American Scientist 48, pp. 160-178, 1960g.

Distribution and motion of interstellar hydrogen in the Galactic System with particular reference to the region within 3 kiloparsecs of the center.
J.H. Oort and G.W. Rougoor, Proc. Nat. Acad. Sci. Washington 46, pp. 1-13, 1960h.

Levensbericht van Walter Baade.
Jaarboek Kon. Ned. Akad. Wetenschappen, 1960/1961, pp. 281-284, 1961a.

Dr. W. Baade.
Hemel en Dampkring 59, pp. 2-17, 1961b.

Dr. Jean Jaques Raimond in memoriam.
Hemel en Dampkring 59, pp. 254-256, 1961c.

A European project for a large observatory in the Southern hemisphere.
International Council of Scientific Unions Rev. 3, pp. 30-35, 1961d

Dynamics of regular galaxies.
In: "Problems of Extragalactic Research", Ed. G.C. McVittie, IAU Symp. 15, pp. 137-145, 1962a.

Normal galaxies and stellar systems : summary.
IAU Symp. 15, pp. 179-183, 1962b.

Presidential Address at the Inaugural Ceremony of the Eleventh General Assembly of the International Astronomical Union, Berkeley 1961.
Trans. I.A.U. 11B, pp. 7-13, 1962c.

The constants of differential galactic rotation.
Trans. I.A.U. 11B, pp. 397-399, 1962d.

Review of problems requiring a resolution of one minute of arc.
Trans I.A.U. 11B, p. 401, 1962e.

Radio data on the distribution and motion of interstellar gas.
In: "The distribution and motion of interstellar matter in galaxies", Ed. L. Woltjer, Benjamin, New York, pp. 3-22, 1962f.

Some remarks on the transition region between disk and halo.
Ibidem, pp. 71-77, 1962g.

Measurement of polarization of the continuous radiation at 75-cm wavelength.
Ibidem, pp. 78-79, 1962h.

Spiral structure.
Ibidem, pp. 234-244, 1962i.

Some considerations concerning the study of the Universe by means of large radio telescopes.
O.E.C.D. Colloquium, Paris; Benelux Cross-Antenna Project Memo 13, 1962j.

Spectrofotometrische waarnemingen van super-reuzen op het Zuidelijk Station der Leidse Sterrewacht door Dr. Th. Walraven en Mevrouw J.H. Walraven.
Verslagen Kon. Ned. Akad. Wetenschappen, Afd. Natuurkunde 72, pp. 10-17, 1963a.

Hydrogène neutre dans la Couronne Galactique?
C.A. Muller, J.H. Oort and E. Raimond, C.R. Acad. Sci. Paris 257, pp. 1661-1662, 1963b.

Empirical data on the origin of comets.
In: "The Moon, Meteorites and Comets", Eds B.M. Middlehurst and G.P. Kuiper (The Solar System, Vol. IV), Univ. Chicago Press, pp. 665-673, 1963c.

Recent large radio telescopes.
Telecommun. J. 30, pp. 313-320, 1963d.

Interstellaire wolken met hoge snelheden.
J.H. Oort, A. Blaauw, A.N.M. Hulsbosch, C.A. Muller, E. Raimond and C.R. Tolbert, Verslagen Kon. Ned. Akad. Wetenschappen, Afd. Natuurkunde 73, pp. 94-101, 1964a.

Structure of the Galaxy.
In: "The Galaxy and the Magellanic Clouds", Eds. F.J. Kerr and A.W. Rodgers, IAU Symp. 20, pp. 1-9, 1964b.

A large high-velocity cloud at $l^{II}=41^{o}$, $b^{II}=-15^{o}$.
IAU Symp. 20, p. 130, 1964c.

Recent observations at Dwingeloo of the central region of the Galactic System.
IAU Symp. 20, pp. 179-183, 1964d.

Stellar Dynamics.
In: "Galactic Structure", Eds. A. Blaauw and M. Schmidt, Univ. Chicago Press, pp. 455-501, 1965a.

Structure and evolution of the Galactic System (Invited discourse at the IAU General Assembly).
Trans. IAU 12A, pp. 789-809, 1965b.

Structure et évolution du système galactique.
L'Astronomie, Bulletin de la Société Astronomique de France 79, pp. 381-389 and 425-444, 1965c.

Aufbau und Entwicklung des Milchstrassensystems.
Erde und Weltall 1, Nr. 2, pp. 14-22, and Nr. 3, pp. 8-17, 1965d.

Bouw en evolutie van het Melkwegstelsel.
Hemel en Dampkring 63, pp. 179-204, 1965e.

Maelkevejssystemets struktur og udvikling.
Nordisk Astronomisk Tidsskrift, pp. 79-105, 1965f.

Struktur und Entwicklung des galaktischen Systems.
Sterne 41, pp. 169-191, 1965g.

Some topics concerning the structure and evolution of galaxies.
In: "The Structure and Evolution of Galaxies", Proc. 13th Conf. Phys. Univ. Brussels (1964), Interscience Publ. London, pp. 17-22, 1965h.

De Leidse Sterrewacht honderd jaar na haar bouw.
In: "Honderd Jaar Sterrewacht", pp. 3-9, 1965i.

High-latitude, high-velocity clouds.
Trans. I.A.U. 12B, pp. 395-397, 1966a.

Possible interpretations of the high-velocity clouds.
B.A.N. 18, pp. 421-438, 1966b.

De grote radiotelescoop in Westerbork.
Mededelingenblad TCHAF, Univ. Leiden 3, pp. 3-12, 1966c.

B. Lindblad.
Quarterly J. Roy. Astr. Soc. 7, pp. 329-341, 1966d.

On the frequency of supernova outbursts in galaxies.
P. Katgert and J.H. Oort, B.A.N. 19, pp. 239-245, 1967a.

E. Hertzsprung, 1873 Oct. 8 - 1967 Oct. 21.
J.H. Oort and K. Gyldenkerne, Nordisk Astronomisk Tidsskrift, pp. 133-138, 1967b.

Possible interpretations of the high-velocity gas.
In: "Radio Astronomy and the Galactic System", Ed. H. van Woerden, IAU Symp. 31, pp. 279-288, 1967c.

Het ontstaan van het Heelal.
Natuur en Techniek 35, pp. 217-224, 1967d.

Waterstofwolken met hoge snelheid en de invanging van gas door het Melkwegstelsel.
Verslagen Kon. Ned. Akad. Wetenschappen, Afd. Natuurkunde 77, pp. 38-43, 1968a.

A new phenomenon in the neighbourhood of the Galactic Centre.
In: "Non-stable phenomena in galaxies", Acad. Sci. Armenian SSR Publ. House, IAU Symp. 29, pp. 41-45, 1968b.

On the possibility of amplification of 21-cm radio emission in high-velocity clouds.
H.G. van Bueren and J.H. Oort, B.A.N. 19, pp. 414-416, 1968c.
(Erratum: B.A.N. 20, p. 224, 1969)

Survey of possible programmes with the SRT at 1415 MHz.
Synthesis Radio Telescope Project, Internal Technical Report 74, 1968d.

Radio-astronomical studies of the Galactic System (Vetlesen Prize Lecture, 1966).
In: "Galaxies and the Universe", Ed. L. Woltjer, Columbia Univ. Press, New York, pp. 1-32, 1968e.

Het extragalactische.
Hemel en Dampkring 67, pp. 63-68, 1969a.

Infall of gas from intergalactic space.
Nature 224, pp. 1158-1163, 1969b.

Spiral structure of galaxies.
In: "Galactic Astronomy", Ed. H.Y. Chiu, Stony Brook, Vol. 1, pp. 121-145, 1970a.

Survey of spiral structure problems.
In: "The Spiral Sructure of our Galaxy", Eds. W. Becker and G. Contopoulos, IAU Symp. 38, pp. 1-5, 1970b.

Matter far from the Galactic Plane associated with spiral arms.
IAU Symp. 38, pp. 142-146, 1970c.

Desiderata for future work.
IAU Symp. 38, pp. 474-478, 1970d.

De vorming van melkwegstelsels in een expanderend heelal.
Verslagen Kon. Ned. Akad. Wetenschappen, Afd. Natuurkunde 79, pp. 24-29, 1970e.

The formation of galaxies and the origin of the high-velocity hydrogen.
Astron. Astrophys. 7, pp. 381-404, 1970f.

The density of the Universe.
Astron. Astrophys. 7, pp. 405-407, 1970g.

Galaxies and the Universe (70th Birthday Symposium Lecture).
Science 170, pp. 1363-1370, 1970h.

De betekenis van de synthese radiotelescoop te Westerbork.
Natuur en Techniek 38, No 6, pp. 9-17, 1970i.

Afscheidscollege (Farewell Lecture, Leiden University; translation in 1971d).
Ned. Tijdschr. Natuurkunde 36, pp. 321-325, 1970j.

The Magellanic Clouds: Summary and desiderata.
In: "The Magellanic Clouds", Ed. A.B. Muller, Reidel, Dordrecht, pp. 184-189, 1971a.

Composition and activity of the nucleus of our Galaxy, and comparison with M31.
In: "Nuclei of Galaxies", Ed. D.J.K. O'Connell, North-Holland Publ. Amsterdam, pp. 321-344, 1971b.

Radiowaarnemingen van kosmische röntgenbronnen met de Synthese Radio Telescoop te Westerbork.
J.H. Oort, L.L.E. Braes, W.N. Brouw and G.K. Miley, Verslagen Kon. Ned. Akad. Wetenschappen, Afd. Natuurkunde 80, pp. 58-61, 1971c.

To the horizon of the Universe.
Delta 14 no. ii, p. 33-45, 1971d.

Absence of radio emission from Maffei I.
Nature 230, pp. 103-105, 1971e.

De Synthese Radio Telescoop in Westerbork en zijn eerste waarnemingen.
In: Z.W.O.-Jaarboek 1970, pp. 141-149, 1971f.

Nieuwe inzichten in spiraalstructuur en explosieve kernen van sterrenstelsels, verkregen met de Synthese Radio Telescoop in Westerbork.
J.H. Oort, R.J. Allen, E. Bajaja, Elly Dekker, P.C. van der Kruit, D.S. Mathewson, A.H. Rots and W.W. Shane, Verslagen Kon. Ned. Akad. Wetenschappen, Afd. Natuurkunde 81, pp. 129-135, 1972a.

The radio emission of NGC 4258 and the possible origin of spiral structure.
P.C. van der Kruit, J.H. Oort and D.S. Mathewson, Astron. Astrophys. 21, pp. 169-184, 1972b. (Erratum: 22, pp. 479, 1973)

De oorsprong van het Heelal.
Holl. Mij. Wetenschappen, Haarlemse Voordrachten 32, pp. 5-27, 1972c.

The development of our insight into the structure of the Galaxy between 1920 and 1940.
In: "Education in and History of Modern Astronomy".
Ed. R. Berendzen, Ann. New York Acad. Sci. 198, pp. 255-266, 1972d.

On the problem of the origin of spiral structure (Karl Schwarzschild Lecture, Wien 1972).
Mitt. Astron. Gesellschaft 32, pp. 15-31, 1973a.

On the problem of the origin of spiral structure.
Commentarii Pontif. Acad. Sci. 2, No. 55, pp. 1-8, 1973b.

G.B. van Albada, 28 Maart 1911 - 18 December 1972.
J.H. Oort and T. de Groot, Hemel en Dampkring 71, pp. 47-48, 1973c.

Note on Verschuur's article on high-velocity clouds and "normal" galactic structure.
A.N.M. Hulsbosch and J.H. Oort, Astron. Astrophys. 22, pp. 153-154, 1973d.

Een afscheid van Professor P.Th. Oosterhoff.
Acta et Agenda, Univ. Leiden, 29 November 1973e.

Recent radio work in nearby galaxies.
In: "Galaxies and Relativistic Astrophysics", (Proc. First Eur. Astr. Meeting, Athens 1972, Vol. 3), Eds. B. Barbanis and J.D. Hadjidemetriou, Springer, Berlin, pp. 1-14, 1974a.

Recent radio studies of bright galaxies.
In: "The Formation and Dynamics of Galaxies", Ed. J. Shakeshaft, IAU Symp. 58, pp. 375-397, 1974b.

The space density of faint M-dwarfs.
Highlights of Astronomy 3, pp. 417-418, 1974c.

The Galactic Centre.
In: "Galactic Radio Astronomy", Eds. F.J. Kerr and S.C. Simonson, IAU Symp. 60, pp. 539-547, 1974d.

Zwakte van Kohoutek was te verwachten.
Zenit 1, No. 2, pp. 13-14, 1974e.

Evolution der Galaxien.
Nova Acta Leopoldina (Abh. D. Akad. Naturf. Leopoldina Halle) 42, pp. 97-112, 1975a.

Phenomenology of spiral galaxies.
In: "Structure and Evolution of Galaxies", Ed. G. Setti, Reidel, Dordrecht, pp. 85-117, 1975b.

Structure of the Hyades cluster.
G. Pels, J.H. Oort and H.A. Pels-Kluyver, In: "Dynamics of Stellar Systems", Ed. A. Hayli, IAU Symp. 69, pp. 159-160, 1975c.

Origine explosive.
In: "La Dynamique des Galaxies Spirales", Ed. L. Weliachew, Coll. Internat. CNRS No. 241, pp. 517-529, 1975d.

Some concluding remarks.
Ibidem, p. 533, 1975e.

Structuren van extra-galactische radiobronnen.
J.H. Oort, H. van der Laan, G.K. Miley and R.G. Strom, Verslagen Kon. Ned. Akad. Wetenschappen, Afd. Natuurkunde 84, pp. 5-13, 1975f

The distance to the Galactic Centre, derived from RR Lyrae variables, the distribution of these variables in the Galaxy's inner region and halo, and a rediscussion of the galactic rotation constants.
J.H. Oort and L. Plaut, Astron. Astrophys. 41, pp. 71-86, 1975g.

New members of the Hyades Cluster and a discussion of its structure.
G. Pels, J.H. Oort and H.A. Pels-Kluyver, Astron. Astrophys. 43, pp. 423-441, 1975h.

Een nieuwe fase in de exploratie van het Heelal.
In: 400 Jaar Leidse Universiteit, pp. 30-35, 1975i.

De kern van ons Melkwegstelsel.
Natuur en Techniek 43, pp. 780-801, 1975j.

Giant radio galaxies.
R.G. Strom, G.K. Miley and J.H. Oort, Scientific American 233, No. 2, pp. 26-35, 1975k.

Levensbericht van Willem Hendrik van den Bos.
Jaarboek Kon. Ned. Akad. Wetenschappen 1975, pp. 208-209, 1976a.

The nucleus of our Galaxy.
Royal Greenwich Observatory Bull. 182, pp. 31-56, 1976b.

Interpretation of radio and optical observations of NGC 1275.
Publ. Astr. Soc. Pacific 88, pp. 591-593, 1976c.

A survey of planetary nebulae near the Galactic Centre.
Publ. Astr. Soc. Pacific 88, pp. 596-597, 1976d.

Over het werk van Rudolph Minkowski.
Zenit 3, pp. 173-174, 1976e.

New observations of the NGC 1275 phenomenon.
V.C. Rubin, W.K. Ford, C.J. Peterson and J.H. Oort, Astrophys. J. 211, pp. 693-696, 1977a.

The expected number density of globular clusters near the Galactic Centre.
Astrophys. J. Letters 218, L97-L101, 1977b.

The Galactic Centre.
Ann. Rev. Astron. Astrophys. 15, pp. 295-362, 1977c.

The Galactic Centre,
Comments Astrophys. 7, pp. 51-66, 1977d.

Introduction to "Structure and Properties of Nearby Galaxies".
In: IAU Symp. 77, Eds. E.M. Berkhuijsen and R. Wielebinski, pp. xvii-xviii, 1978a.

Obituary for Prof. P.Th. Oosterhoff.
IAU Information Bulletin 40, 1978b.

Evolutie in het Heelal.
Verslagen Kon. Ned. Akad. Wetenschappen, Afd. Natuurkunde 87, pp. 120-125, 1978c.

Speculations on the origin of the Chain A of high-velocity clouds.
In: "Problems of Physics and Evolution of the Universe", Ed. L.V. Mirzoyan, Acad. Sci. Armenian SSR, pp. 259-280, 1978d.

Terugzien in de tijd.
Natuur en Techniek 46, pp. VI-XIII, 1978e.

Eruptive phenomena near the Galactic Centre.
Physica Scripta 17, pp. 175-184, 1978f.

Remark on activity in nuclei of spiral galaxies different from that in Seyfert galaxies.
Physica Scripta 17, p. 369, 1978g.

High-velocity clouds.
J.H. Oort and A.N.M. Hulsbosch, in: "Astronomical Papers dedicated to Bengt Strömgren", Eds. A. Reiz and T. Andersen, Copenhagen Univ. Obs., pp. 409-428, 1978h.

The galactic distribution of OH/IR stars.
B. Baud, H.J. Habing and J.H. Oort, in: "The Large-Scale Characteristics of the Galaxy", Ed. W.B. Burton, IAU Symp. 84, pp. 29-34, 1979a.

Introduction to the session on the Galactic Nucleus.
IAU Symp. 84, p. 323, 1979b.

Great Expectations.
ESO Messenger 16, pp. 17-19, 1979c.

Bannier's betrokkenheid bij dertig jaar sterrenkunde.
In: "Onder de 'ZWO-Bannier'", ZWO, Den Haag, pp. 85-101, 1979d.

Luminosity distribution and shape of the Hyades Cluster.
Astron. Astrophys. 78, pp. 312-317, 1979e.

Non-standard abbreviations:
B.A.N. = Bull. Astron. Inst. Netherlands.
IAU = Internat. Astron. Union.

This bibliography was compiled by Dini Ondei, Jay Ekers, Willem N. Brouw and Hugo van Woerden.

Dini Ondei studied English Literature at Leiden, and has been Professor Oort's secretary since about 1962.

Jay Ekers studied chemistry at Adelaide, and has been a research assistant at Radiophysics CSIRO, California Institute of Technology, and since 1975 at Groningen (Netherlands Foundation for Radio Astronomy).

Willem N. Brouw studied astronomy at Leiden (1957-63) and obtained his doctorate there in 1971. He is head of the Computer Group in the Netherlands Foundation for Radio Astronomy and part-time Professor at Leiden.

Hugo van Woerden studied astronomy at Leiden (1945-55) and obtained his doctorate at Groningen in 1962. He has been at the Kapteyn Institute since 1957, and has been Professor of Radio Astronomy at Groningen since 1965.

SUBJECT INDEX TO OORT'S BIBLIOGRAPHY

This index was compiled by Hugo van Woerden. It is essentially based only on the titles of papers, not on Abstracts or any more detailed analysis.

GENERAL SUBJECT INDEX

Compiled by Hugo van Woerden.

NAME INDEX

Underlined numbers refer to the first page of papers in the present volume contributed by the person in question.

Oort, J.H. has not been indexed.